HUMAN ORIGINS

The Fossil Record
Third Edition

HUMAN ORIGINS
The Fossil Record
Third Edition

Clark Spencer Larsen
University of North Carolina, Chapel Hill

Robert M. Matter
Sauk Valley College

Daniel L. Gebo
Northern Illinois University

WAVELAND
PRESS, INC.
Prospect Heights, Illinois

For information about this book, write or call:
 Waveland Press, Inc.
 P.O. Box 400
 Prospect Heights, Illinois 60070
 (847) 634-0081

Cover design: Robert and Sara Matter
Cover illustration: Australopithecus afarensis AL444-2

ISBN 1-57766-002-1

Printed in the United States of America

7 6 5 4 3 2

Table of Contents

THE AUTHORS

Clark Spencer Larsen has taught at the University of Massachusetts (North Dartmouth campus), Northern Illinois University, and Purdue University. He is currently Professor of Anthropology in the Department of Anthropology at the University of North Carolina, Chapel Hill. He has an appointment as Research Associate in the Research Laboratories of Archaeology at the University of North Carolina and in the Department of Anthropology at the American Museum of Natural History, and is Adjunct Professor of Biological Anthropology and Anatomy at Duke University. He is the President-Elect of the American Association of Physical Anthropologists. He received his B.A. in anthropology from Kansas State University, and both his M.A. and Ph.D. are in biological anthropology from the University of Michigan. His research is primarily focussed on bioarchaeology and biocultural adaptation in Holocene human populations. Although his fieldwork and related research is mostly in the southeastern United States, he has been involved in projects in the American Great Basin, New England, Midwest, and Plains, and has studied the skeletons of Neandertals from Krapina, Croatia, and Neolithic farmers from Greece. He has authored, edited, and co-edited over a dozen books and monographs, including *Advances in Dental Anthropology* (Wiley-Liss, New York), *In the Wake of Contact: Biological Responses to Conquest* (Wiley-Liss, New York), and *Bioarchaeology: Interpreting Behavior from the Human Skeleton* (Cambridge University Press, Cambridge). His articles and other works have appeared in *Current Anthropology, Journal of Human Evolution, American Journal of Physical Anthropology, Human Biology, International Journal of Osteoarchaeology, Advances in Archaeological Method and Theory,* and *Anthropological Papers of the American Museum of Natural History.*

Robert M. Matter is Professor Emeritus of Art and Anthropology at Sauk Valley College (Dixon, Illinois). After studying art at the School of the Art Institute of Chicago, he continued in the College of Visual and Performing Arts at Northern Illinois University where he received his B.A. and M.A. degrees. He also received an M.A. in anthropology from that institution.

Daniel L. Gebo is Professor of Anthropology at Northern Illinois University (DeKalb, Illinois). He received his B.S. in zoology from Oregon State University; his M.S. in anthropology from the University of Wisconsin, Milwaukee; and his Ph.D. in biological anthropology from Duke University. His research focusses on primate evolution, with a particular emphasis on locomotor adaptation, functional morphology, and phylogeny. His field research has taken him to the badlands of Wyoming, Uganda, and China, the deserts of Egypt, Mali, and Colombia, and the rain forests of Costa Rica, Uganda, Madagascar, and the Philippines. Results of his research have been published in the *Journal of Human Evolution, American Journal of Physical Anthropology, Folia Primatologica, American Journal of Primatology, Primates, Science,* and *Nature.* He also edited the book *Postcranial Adaptation in Non-human Primates* (Northern Illinois University Press, DeKalb).

PREFACE

In the following pages, we present an introductory guide to the fossil record of prehuman and human evolution through the presentation of detailed drawings of complete or nearly complete specimens that are representative of particular grades of evolution. Each drawing is accompanied by appropriate information, including geographic location, approximate age, and general description. The age of each of the specimens is expressed with reference to the geological epoch as well as an approximate absolute date in millions of years before the present (myBP) or years before the present (yBP) for the more recent materials. The specimen descriptions include information on the context of the discovery and summary anatomical details. Additionally, we have provided references that will give further documentation and discussion of individual fossil specimens. The references include a mix of technical and popular literature. Our classroom experience has shown that these references provide a useful tool for directing students to key literature, especially with regard to preparation of term papers, reports, and other activities.

The volume is organized chronologically beginning with common ancestors of apes and humans, followed by proposed hominid ancestors, continuing with the first hominids, and tracing the development of the pre-australopithecine, australopithecine, and *Homo* lineages through the evolution of modern *Homo sapiens*. The fossils are grouped by continent in order to show regional evolutionary trends. If more than one individual has been recovered from a fossil locality, the specimens that are illustrated in the text are numbered or, in some instances, lettered following the system presented in the *Catalogue of Fossil Hominids* (e.g., Sangiran 4).

More often than not, it is the hardest of bones—usually cranial bones—and teeth that survive in the ground. The collection of remains, therefore, that we have chosen to discuss in this book is primarily skulls and teeth. In order to facilitate comparisons of specimens presented in this book, we have chosen to illustrate most of them at 75% of original size. The exceptions include the following: all occlusal views of dentitions (100%), M.N.H.N.P. A.C. 36 mandible (100%), *G. blacki* III mandible (100%), A.L. 333-63 phalanx (100%), *H. sapiens* phalanx (100%), O.H. 8 partial foot (100%), IGF 11778 skeleton (19%), A.L. 288-1 skeleton (19%), KNM-WT 15000 skeleton (16%), and STS 14 skeleton (50%).

Other guides to the fossil record are available, including *Catalogue of Fossil Hominids* (Oakley, Campbell, and Molleson, 1971, 1975, 1977) and *Guide to Fossil Man* (Day, 1986). These sources provide far more information than is necessary for the undergraduate student grappling with—usually for the first time—an increasingly complex fossil record. Several introductory summaries with illustrations are available (Brace, Nelson, Korn, and Brace, 1979; Sauer and Phenice, 1977), including the first and second editions of the present book (Larsen and Matter, 1985, 1991), but because more materials have been recovered by paleontologists in the intervening years since the publication of these works, we felt compelled to prepare an updated guide.

We do not pretend to provide here an exhaustive compilation of the fossil record of hominoid evolution. Rather, we have chosen examples that present salient details of evolutionary change. As such, the volume is intended as a supplementary text to be used in conjunction with textbooks written for either the beginning level of study (e.g., Poirier, 1993; Campbell and Loy, 1996) or for the more advanced level (e.g., Klein, 1989; Conroy, 1997; Wolpoff, 1998). We have also strived to include as much of the new material that has been recovered or reported on in the last several years. Most texts on human evolution end with developments that took place during the late Pleistocene and infer or state outright that evolutionary change comes to a grinding halt at that point. We have provided a series of specimens that demonstrate this not to be the case, particularly with respect to biological change consequent to the transition from foraging to food production and agriculture during the Holocene.

We were fortunate in being able to guide the development of many of the illustrations by having direct access to the original specimens housed in a temporary exhibit ("Ancestors") at the American Museum of Natural History in 1984 (Tattersall and Delson, 1984; Lewin, 1984; Pfeiffer, 1984; Delson, 1985). Specimens at the exhibit that we were able to study in some detail include the GSP 15000 face, Taung skull and endocranial cast, STS 5 cranium, STS 71 cranium, SK 23 mandible, SK 48 cranium, StW 53 cranium and cranial reconstruction, Sangiran 4 palate, Mauer 1 mandible, Salé, cranium, Arago 21 face, Steinheim 1 cranium, Amud 1 cranium, Skhul 5 cranium, Krapina 3 cranium, La Ferrassie 1 cranium, La Quina 5 cranium, Saccopastore 1 cranium, Neandertal cranium, and Cerro Sota 2 cranium. We will be forever grateful for the opportunity afforded us in viewing these fossil materials at the American Museum. It is very unlikely that the chance to examine all of these paleontological treasures in one place will occur again. Indeed, we owe the organizers of the exhibit, Ian Tattersall (American Museum of Natural History) and Eric Delson (Herbert H. Lehman College and American Museum of Natural History) a great debt of gratitude for their efforts in bringing these materials to the viewing attention of scientist and nonscientist alike.

Most of the drawings were prepared from the combined use of photographs and cast reproductions of the fossils. In the preparation of the drawings, close attention was given to size details. In this regard, the late Harry L. Shapiro very kindly provided unpublished data on the Cerro Sota 2 modern *Homo sapiens* cranium; dimensions for the other figured specimens were taken from the literature and from casts. Russell H. Tuttle and Geoffrey G. Pope allowed us access to casts of the STS 71 australopithecine cranium and the Sangiran 17 *Homo erectus* cranium, respectively. William H. Kimbel provided us access to a cast of a proximal hand phalanx from Hadar (A.L. 333-63). David Arter gave access to his cast reconstruction of the A.L. 288-1 ("Lucy") skull. Alan Walker permitted us to examine casts of the KNM-WT 17000 australopithecine cranium ("Black Skull") and the KNM-WT 15000 *Homo erectus* juvenile skeleton. High quality photographs of individual fossil specimens were kindly provided by Alan Walker (KNM-WT 15000), Donald Tyler (Sangiran IX), and Kenneth Kennedy (Narmada). Sheilagh T. Brooks allowed access to the study of the cranium from the Humboldt Sink in the collections of the University of Nevada, Las Vegas.

Dennis O'Brien provided advice regarding the production of the figures. Ms. Frances Keefer prepared the maps showing the site locations of the materials figured in this volume. We thank the many people who provided comments on the first and second editions of the book and those who suggested changes for the preparation of the present (third) edition. For their help in this regard, we especially thank Christopher B. Ruff (Johns Hopkins University), Alan C. Walker and Pat Shipman (Pennsylvania State University), Tim D. White (University of California, Berkeley), Fred H. Smith (Northern Illinois University), Scott W. Simpson (Case Western Reserve University), Katherine F. Russell (University of Massachusetts, Dartmouth campus), Hermann S. Helmuth (Trent University), William H. Kimbel (Institute of Human Origins and Arizona State University), Lynne A. Schepartz (University of Cinncinati), Kenneth A.R. Kennedy (Cornell University), and F. D. Burton (University of Toronto, Scarborough). In the preparation of the present edition, Steven E. Churchill (Duke University) provided many valuable comments for improving the clarity and organization of the text. We thank Fabiana Frascaroli (Buenos Aires, Argentina) for providing information on AMS ^{14}C dates of the Cerro Sota 2 cranium from Argentina.

Students are sometimes left with the impression that paleoanthropologists spend their lives looking at dry old bones and fossils. It is important to emphasize that it is these bones and fossils represent once-living animals. As one paleoanthropologist has stated, "We must imagine ourselves looking at these animals as though they were alive today and then ask what would give us critical understanding of them as functioning, living creatures" (Pilbeam, 1984a: 14). It's this dynamic perspective that paleoanthropologists offer which helps to make this such an exciting field. Toward this objective, the beginning student must understand the data base—the fossil record—from which scientists provide interpretations of behavioral and evolutionary history. We believe that the use of visual materials for the student who has little familiarity with this record is essential—we hope that our descriptions and accompanying drawings will fill that need. To this end, we thank our publisher at Waveland Press, Neil Rowe, for his encouragement and support in the preparation of the first, second, and this newly revised edition of *Human Origins: The Fossil Record.*

C.S.L., R.M.M., and D.L.G.

1. INTRODUCTION

a. Paleoecology

An understanding of evolution from the perspective of the ancient past is a complicated undertaking at all levels of study. At every paleontological site, there is information to be learned about the past, especially concerning the environment in which hominids and their predecessors lived. Survival techniques and general lifestyle can be discovered by the use of a complex array of disciplines (e.g., geology, paleontology, and archaeology) which all help to reconstruct a paleoecological context for prehuman and human evolution. Much of the mystery of past populations is due in a large part to the fact that the fossilized and nonfossilized remains that are recovered by paleontologists and other scientists represent only a very small portion of the once living populations from which they are taken. Consequently, many of the discussions by paleoanthropologists often center around the larger issue of what exactly it is that these fossils represent. That is, can we safely say that individual fossils are representative of past populations, and, if so, how? The study of natural processes—taphonomy—associated with the preservation or destruction of past life forms helps us to provide answers to this question by examining in greater detail the process of fossilization and events that take place following the death of hominids and other organisms.

References: Shipman, 1981; Behrensmeyer, 1984; Behrensmeyer and Hill, 1980; Weigelt, 1989.

b. Bone Chemistry and Fossilization

It is usually only the hard tissues—bones and teeth—that remain for anthropologists to study. The hardness of bones and teeth is due to the presence of inorganic minerals. In bone, in particular, the inorganic component accounts for about two-thirds the weight. This includes roughly 85% calcium phosphate and 10% calcium carbonate. The remaining one-third is the organic component, mostly formed by the protein collagen. Fossil bones and teeth are usually harder than more recent bones and teeth. For the most part, this hardness arises from the chemical replacement of the organic and inorganic components of bone by elements in the surrounding matrix in which the remains rest. In general, the older the fossil, the less of the original constituents remain present. The rate of replacement of the original hard tissues is dependent upon local conditions, including water content, minerals in the soil matrix, and a host of other factors. In some instances, very old bones and teeth show relatively little replacement, whereas materials of recent age are known to be well on the way to being fossilized.

References: McLean and Urist, 1968; Goldberg, 1982; Wolpoff, 1998.

c. The Skeleton

The skeleton is divided into two major groups of bones. The *axial* group consists of bones that form the body cavities and act to protect vital organs. They include the skull (cranium and mandible), hyoid (small bone in the neck region), vertebral column, ribs, and sternum. The *appendicular* group includes the limbs—the bones of the shoulder (clavicle and scapula), arm (humerus, radius, and ulna), wrist and hand, pelvis (innominates), leg (femur, patella, fibula, tibia), ankle, and foot. In total, there are 206 bones in the human skeleton.

The human dentition is comprised of two age-successive generations of teeth, the earlier deciduous (or milk) dentition (20 teeth: 8 incisors, 4 canines, and 8 molars) and a later permanent (or adult) dentition (32 teeth: 8 incisors, 4 canines, 8 premolars, and 12 molars). Unlike the skeletal tissues, once the teeth are formed they do not change in morphology except by disease (e.g., decay) or wear. Teeth are perhaps the most studied of the hard tissues that are preserved in the paleontological record. Given their compact size and durability, they have survived much better in the fossil record than bones, thereby providing a greater amount of information about phylogenetic relationships, evolutionary history, and dietary adaptations.

The original skeletal tissues from which fossils originate are from a system that in life serve a number of important functions, including protection of the soft and vital organs of the body (e.g., brain, heart, lungs, liver, kidneys); attachment areas for muscles, ligaments, and tendons; production of red blood cells; and storage of important salts and minerals for later use by the body. The skeletal apparatus also acts as a superstructure for the support of other tissues. Mechanically, bones are comprised of materials and are shaped so as not to deform appreciably under loading. Animals—including humans—engage themselves in a variety of physical behaviors that influence the degree and kind of mechanical loads to which they are subjected. By studying the materials and structure of bones, paleoanthropologists are able to reconstruct some of these behaviors in humans and human ancestors, such as those associated with the acquisition and processing of food, locomotion, general level of activity and pattern of activity, and related behaviors.

References: McLean and Urist, 1968; Goldberg, 1982; Frankel and Nordin, 1980; Currey, 1984; Hildebrand, Bramble, Liem, and Wake, 1985; Shipman, Walker, and Bichell, 1985; Cartmill, Hylander, and Shafland, 1987; Steele and Bramblett, 1988; Fleagle, 1988; White, 1991; Bass, 1995; Schwartz, 1995; Hillson, 1996; Larsen, 1997.

Modern Human Skeleton

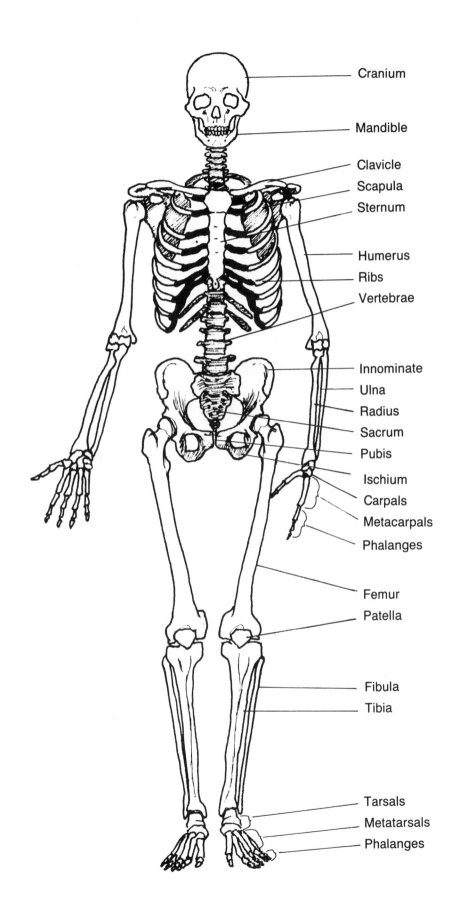

Cranium

Mandible

Clavicle

Scapula

Sternum

Humerus

Ribs

Vertebrae

Innominate

Ulna

Radius

Sacrum

Pubis

Ischium

Carpals

Metacarpals

Phalanges

Femur

Patella

Fibula

Tibia

Tarsals

Metatarsals

Phalanges

Modern Human Skull: Frontal View

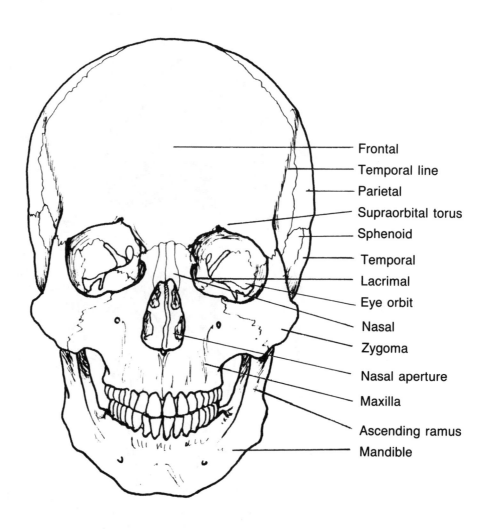

Frontal
Temporal line
Parietal
Supraorbital torus
Sphenoid
Temporal
Lacrimal
Eye orbit
Nasal
Zygoma
Nasal aperture
Maxilla
Ascending ramus
Mandible

Modern Human Skull: Lateral View

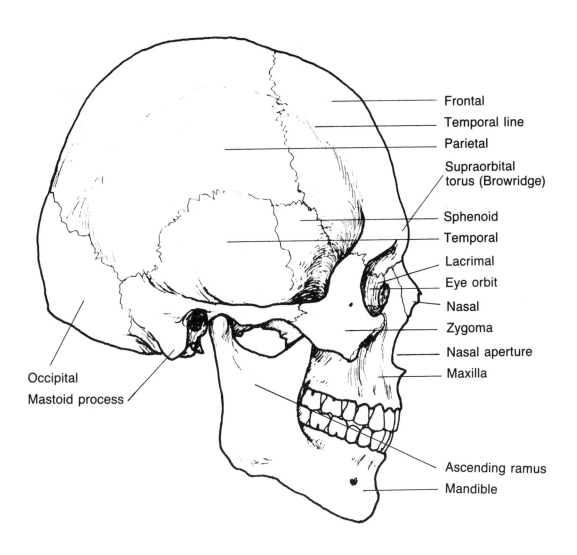

Frontal

Temporal line

Parietal

Supraorbital
torus (Browridge)

Sphenoid

Temporal

Lacrimal

Eye orbit

Nasal

Zygoma

Nasal aperture

Maxilla

Occipital

Mastoid process

Ascending ramus

Mandible

Modern Human Dentition

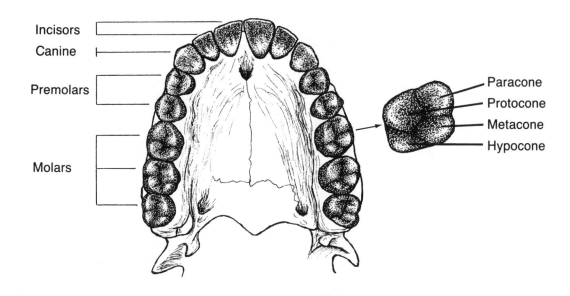

Incisors
Canine
Premolars
Molars

Paracone
Protocone
Metacone
Hypocone

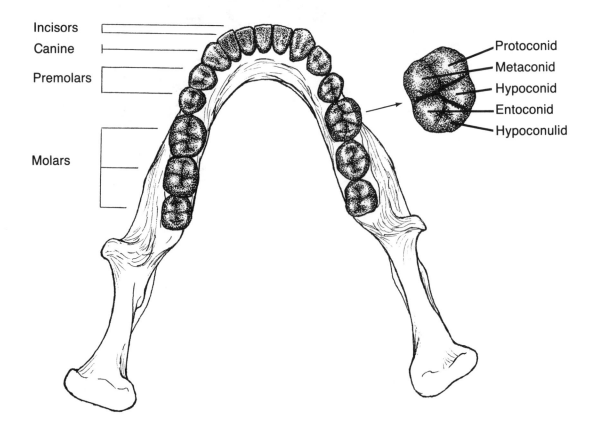

Incisors
Canine
Premolars
Molars

Protoconid
Metaconid
Hypoconid
Entoconid
Hypoconulid

d. Comparative Anatomy

Well before the present century, anatomists and other biologists appreciated the striking morphological similarities between the Great Apes—chimpanzees, gorillas, and orangutans—and humans. An understanding of the anatomy of the skull and dentition of our closest relatives, the Great Apes, provides us with important information with regard to the evolutionary history of the human family, the Hominidae.

The two African Great Apes, the gorilla (*Gorilla*) and chimpanzee (*Pan*), and the single Asian Great Ape, the orangutan (*Pongo*), share a number of anatomical features, but also differ in a variety of ways. For example, in the chimpanzee and gorilla, the eye orbits and supraorbitals (browridges) are large (although see the figure of the pygmy chimpanzee) and the orbits of the orangutan are small. Compared to chimpanzees and gorillas, the eye orbits of the orangutan are small and ovoid-shaped and much narrower than either the gorilla or chimpanzee. The supraorbitals in the orangutan are somewhat smaller than in the other great apes. All the great apes are similar in the anatomical region immediately above and behind the eye orbits—the region is constricted and narrow before the braincase expands outward. This feature is referred to as postorbital constriction.

The male gorilla and male orangutan have large crests of bone that are oriented along the midline, front to back, on the top of the cranium. This crest—called the sagittal crest—is present because of the very large chewing muscles (temporalis) that attach on the sides of the cranium. These muscles act in conjunction with the set of chewing muscles (masseter) that are attached to the cheeks and base and sides of the mandible. Similarly, on the occipital bone, the males of these two genera of apes have large crests—occipital (or nuchal) crests—that run perpendicular to the sagittal crest. The presence of the occipital crest is due to the great development of the neck muscles that attach to this area of the cranium. Occasionally, these crests have been reported in large female gorillas and female orangutans as well as in chimpanzees.

The teeth of pongids show a wide range of morphological variability that is similar to that of hominid teeth. In pongids, the incisors are broad and spatulate. In the maxillary incisors, the lateral incisors tend to be much reduced in size relative to the central incisors. This condition is especially apparent in the orangutan. Almost always, there is a pronounced diastema or gap between the upper lateral incisor and upper canine. This diastema accommodates the mandibular canine which in pongids is large, projecting, and tusk-like. In the lower jaw, there is a diastema between the lower canine and the lower first premolar. In hominids, there is no diastema in either the upper or lower jaw except with regard to those that have been documented in the earliest hominids. Moreover, hominid canines—upper or lower—are small and generally do not project beyond the other teeth.

Ape premolars are also distinctive in their morphology. Upper premolars and the lower second premolars have two cusps each. On the lower first premolar, there are also two cusps, but the cusp on the cheek side of the tooth is considerably larger than the cusp on the tongue side of the tooth. The dominant cusp on the cheek side of the lower first premolar is sectorial and acts as a hone to sharpen the crest on the back of the upper canine. Hominids do not possess a sectorial premolar. Pongid upper molars and hominid upper molars are similar in anatomical detail. In both pongids and hominids, the upper molars have four cusps, including two cusps on the cheek side of the tooth (paracone in front, metacone behind) and two cusps on the tongue side (protocone in front, hypocone behind). Pongid and hominid lower molars are also quite similar in overall morphology. The lower molars in both frequently have five cusps, three on the cheek side (front to back: protoconid, hypoconid, hypoconulid) and two cusps on the tongue side (metaconid in front, entoconid behind). The pattern formed by the separation of the cusps by grooves gives the lower molars a characteristic "Y" shape and is called the "Y-5" dental pattern. In hominids, one or more of these cusps is frequently missing, thus giving the tooth a simpler morphology.

References: Huxley, 1863; Gregory, 1922, 1950; Gregory and Hellman, 1926; Peyer, 1968; Schultz, 1969; Dahlberg, 1971; Le Gros Clark, 1971; Swindler and Wood, 1973; Swindler, 1976; Johanson, 1979; Ankel-Simons, 1983; Fleagle, 1988; Aiello and Dean, 1990.

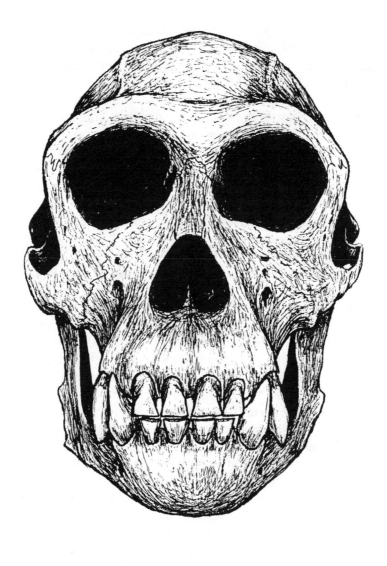

0   cms   5

Chimpanzee, Male

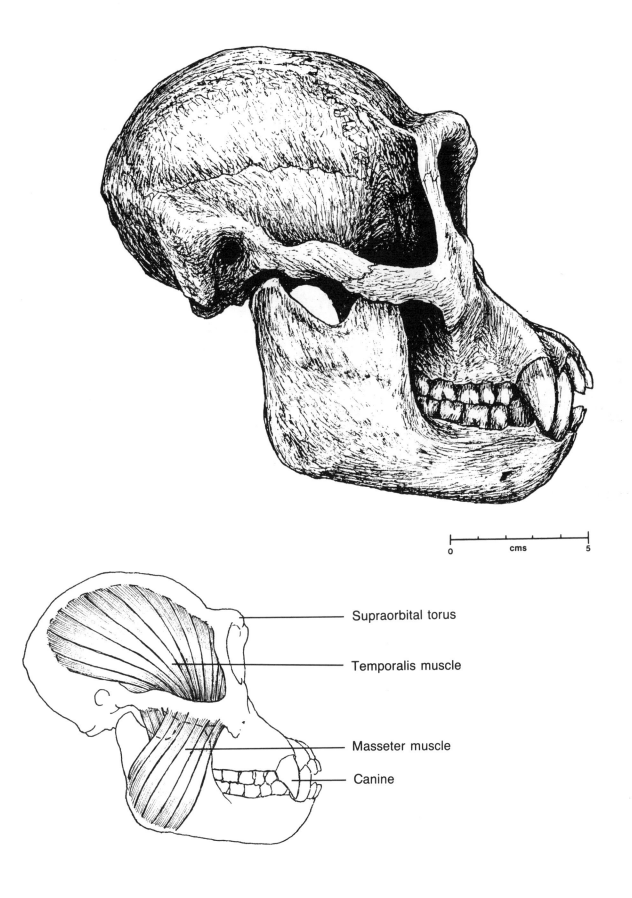

Supraorbital torus

Temporalis muscle

Masseter muscle

Canine

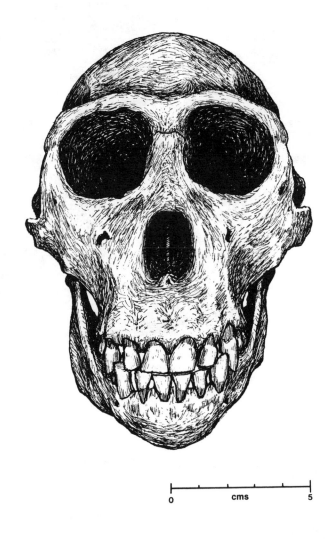

0   cms   5

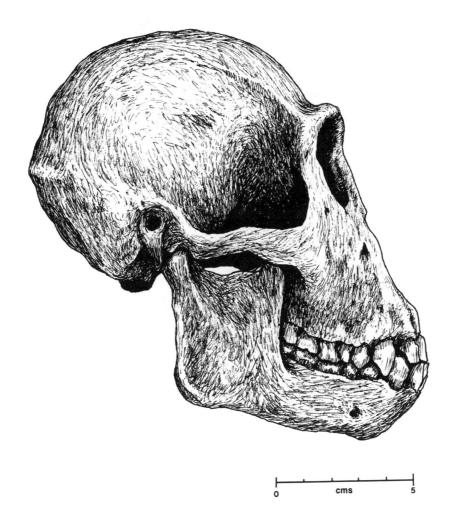

0   cms   5

Pygmy Chimpanzee Dentition, Male

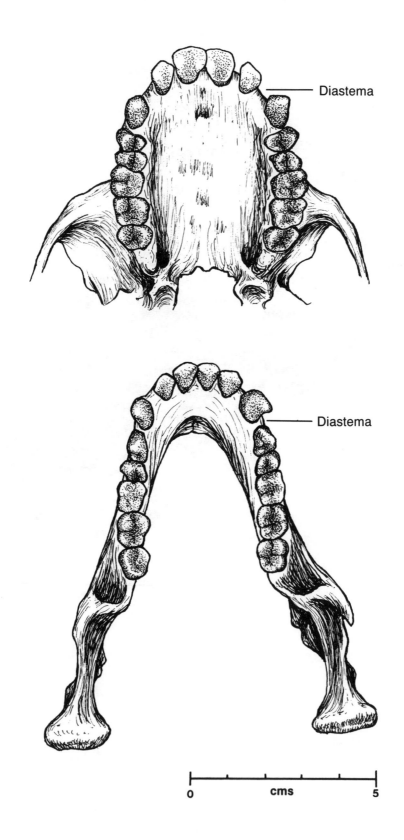

Diastema

Diastema

0   cms   5

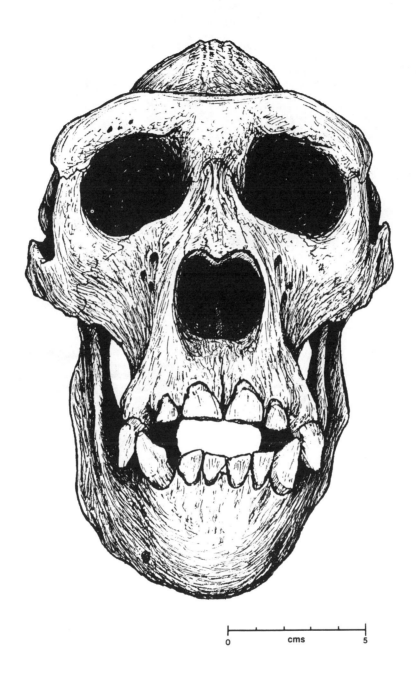

0   cms   5

Gorilla, Female

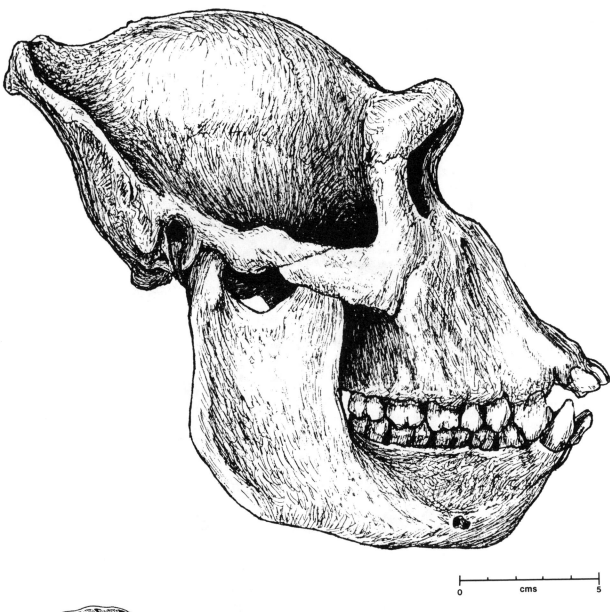

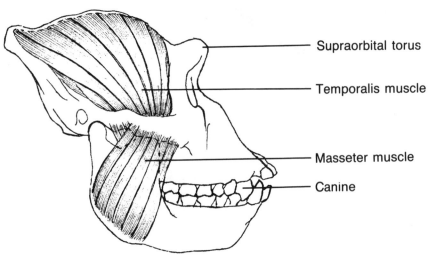

Supraorbital torus

Temporalis muscle

Masseter muscle

Canine

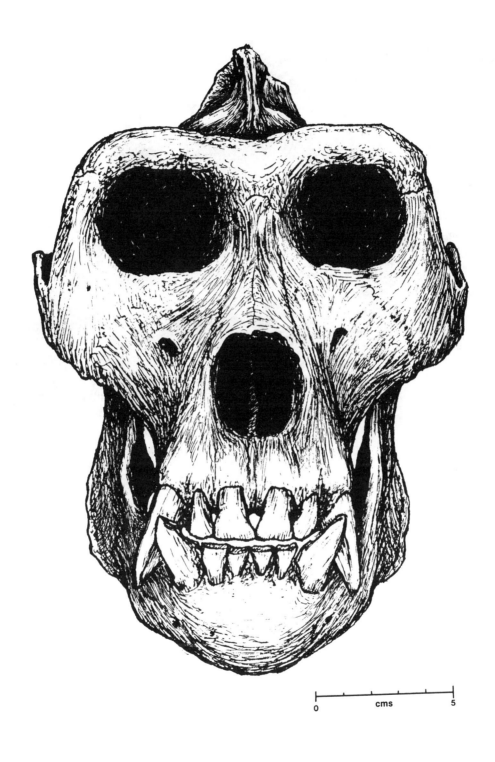

Gorilla, Male

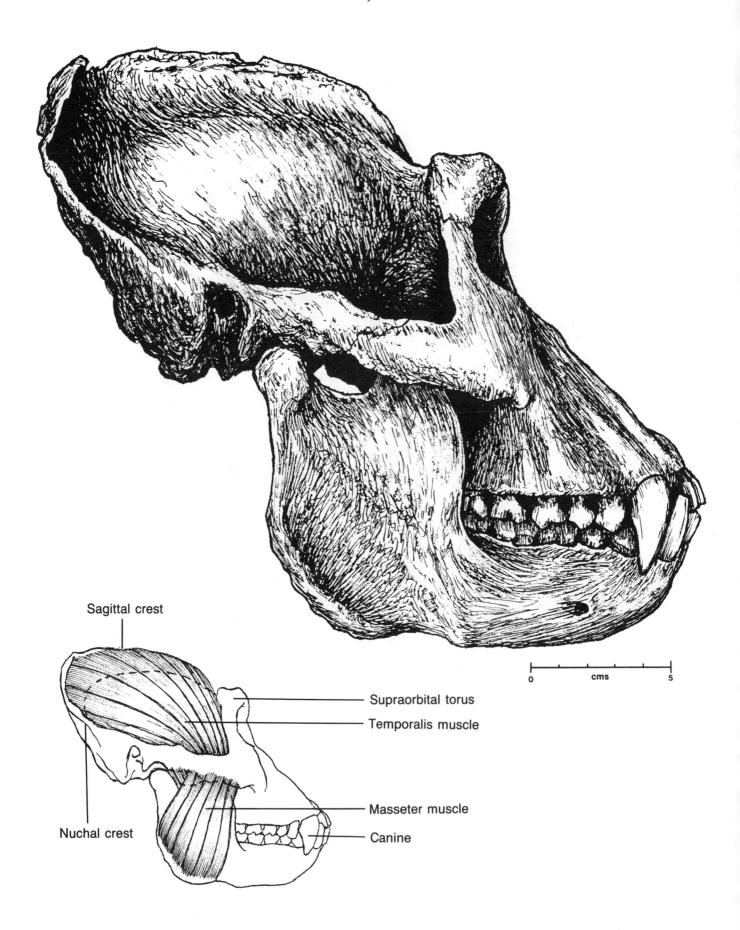

Sagittal crest

Supraorbital torus

Temporalis muscle

Masseter muscle

Canine

Nuchal crest

0   cms   5

Gorilla Dentition, Female

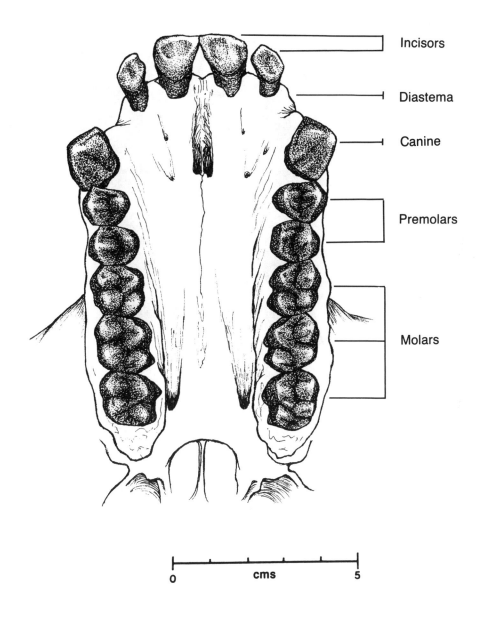

Incisors

Diastema

Canine

Premolars

Molars

0   cms   5

Gorilla Dentition, Female

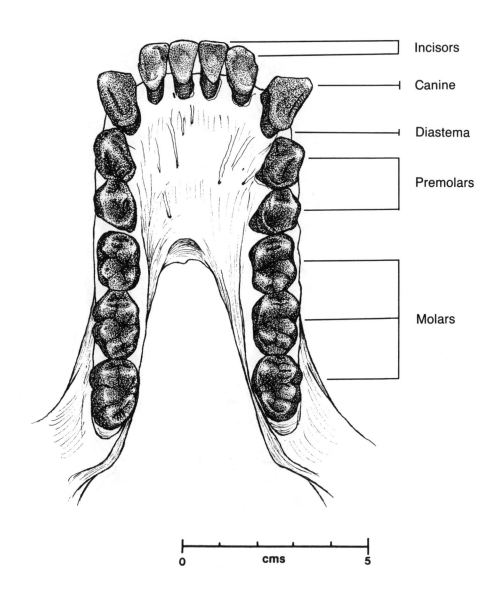

Incisors

Canine

Diastema

Premolars

Molars

0   cms   5

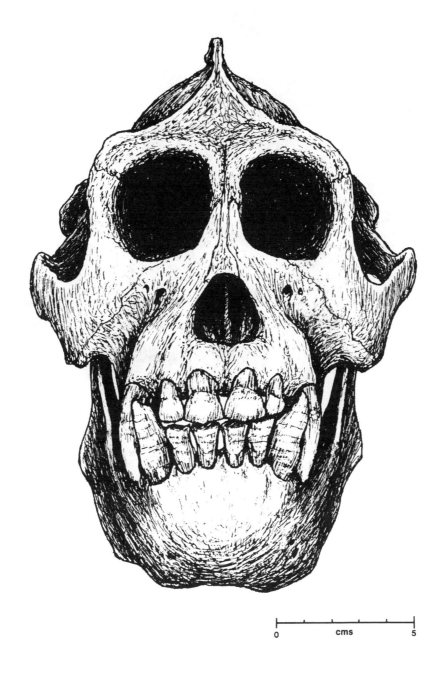

0   cms   5

Orangutan, Male

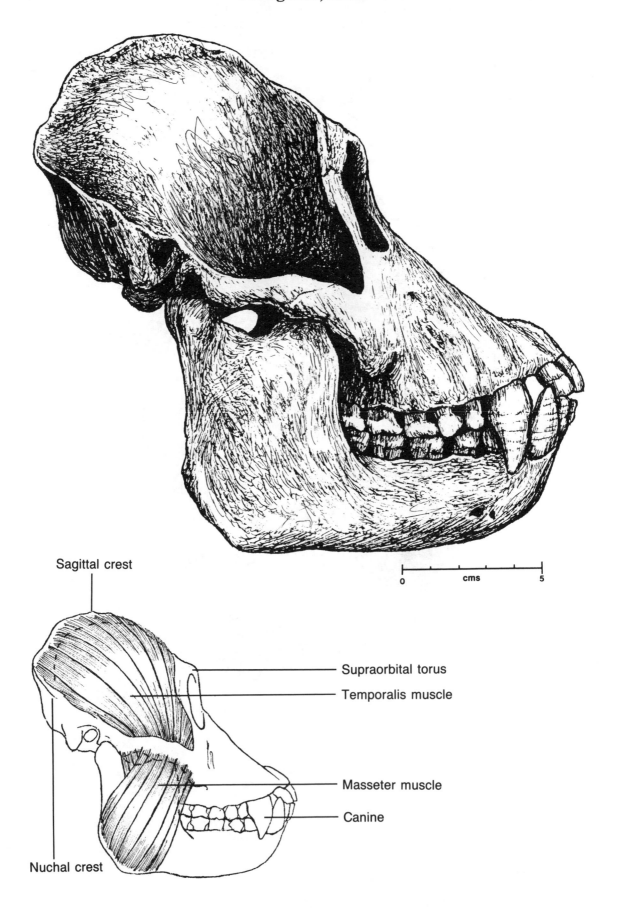

Sagittal crest

Supraorbital torus

Temporalis muscle

Masseter muscle

Canine

Nuchal crest

Dawn Apes

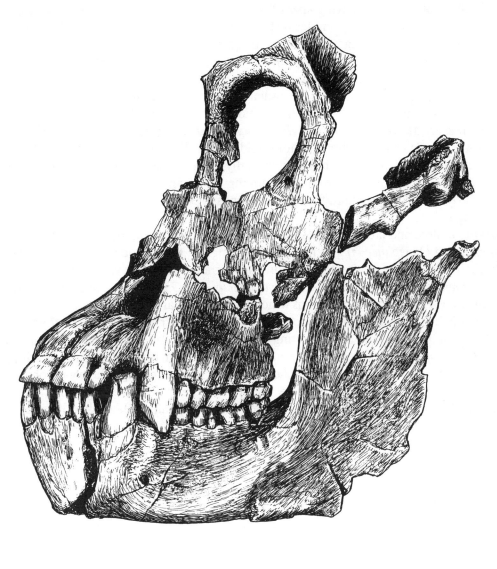

0   cms   5

GSP 15000

2. DAWN APES

Oligocene and Miocene Hominoids

In the early Oligocene (35 myBP), there is evidence for the initial divergence of a primate lineage which eventually evolved into the living apes of today (gibbons, orangutans, chimpanzees, and gorillas). As might be expected, these early primates lack many of the key adaptations which are typically associated with the living apes (e.g., long forelimbs for frequent climbing or brachiation). In the Oligocene of Africa, one primate in particular—*Aegyptopithecus*—represents a very primitive form compared to later hominoids in the Miocene and especially to the living apes.

By the early Miocene (22.5 myBP), the radiation of hominoids was extremely diverse in terms of number of species and in body sizes, but all hominoids are still geographically located within Africa. *Proconsul, Afropithecus, Morotopithecus,* and *Turkanapithecus* represent four genera of this diverse and broadly adapted group of early Miocene hominoids. Dentally, these early Miocene hominoids share many features with the living apes, but postcranially, they are still remarkably primitive and lack a number of limb specializations known to occur among the living apes, with one key exception. *Morotopithecus*, an early Miocene hominoid from Uganda, possesses a lower back and shoulder anatomy like that of living apes.

Toward the middle and late Miocene, hominoids continued to undergo adaptive radiation in Africa (*Kenyapithecus* and *Otavipithecus*) and had begun to migrate to Europe and Asia. Non-African genera that make their appearance during this time include, among others, *Dryopithecus* and *Oreopithecus* in Europe and *Lufengpithecus, Ankarapithecus, Sivapithecus,* and *Gigantopithecus* in Asia. *Dryopithecus* and *Oreopithecus* show specific limb adaptations which are very similar to the limb anatomy of the living apes. *Lufengpithecus* is less well known but appears to be more similar to *Dryopithecus* than to *Sivapithecus,* whereas *Ankarapithecus* displays an interesting array of modern and primitive ape facial features. *Sivapithecus* and *Gigantopithecus* represent a dietary shift from other frugivorous (fruit-eating) hominoids. Both genera show great similarity in their chewing apparatus reflecting a heavy emphasis on crushing and grinding of food. The mandibles are very robust, and the molars are large and have thick enamel. This dietary shift is similar to what is later observed in the hominid lineage as well as the mid-Miocene African hominoid called *Kenyapithecus*, a potential ancestor for African apes and humans.

The facing map shows the site locations of fossils representing these hominoids:

1. Fayum, Egypt: DPC-2803, DPC-1028
2. Rusinga Island, Kenya: KNM-RU 7290
3. West Turkana, Kenya: KNM-WK 16999, KNM-WK 16950A
4. Moroto, Uganda: UM.P. 62-11
5. St. Gaudens, France: M.N.H.N.P.-A.C. 36
6. Monte Bamboli, Italy: IGF 11778
7. Yassiören, Turkey: AS95-500
8. Lufeng, China: IVPP PA580, IVPP 644
9. Potwar Plateau, Pakistan: GSP 15000
10. Kwangsi, China: *Gigantopithecus blacki* Mandibles II and III
11. Maboko Island, Kenya: KNM-MB 20573
12. Berg Aukas, Namibia: BER I, 1'91

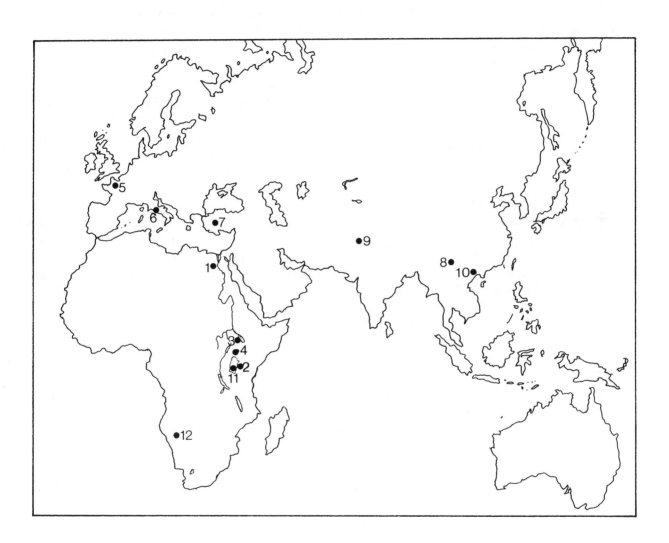

DPC-2803, DPC-1028

Specimen: **DPC-2803, DPC-1028**
Geographic location: Fayum, Egypt

Taxonomic affiliation: *Aegyptopithecus zeuxis*
Dating: Early Oligocene (>31 myBP)

General Description: Due to the efforts of Elwyn Simons and his colleagues, a plethora of specimens of early Oligocene primates have been discovered over the last three decades from the eastern Sahara desert in Egypt from the locality known as the Fayum. These fossils consist of cranial materials, including facial specimen DPC-2803 and mandible DPC-1028 shown on the facing page, and numerous postcrania. The cranial and facial anatomy of this Old World primate document some of the earliest evidence of complete postorbital closure, a fused frontal bone, and advanced brain complexity—all of which support the anthropoid status given this fossil primate. Similarly, note the prominent temporal crests which pinch together above the browridges and the massive development of the cheek region for attachment of chewing musculature. Both features are more commonly associated with living apes. Its small orbits indicate a diurnal (daytime) lifestyle. DPC-2803 is believed to be an old adult male due to the large canine sockets and the prominent muscle crests on this specimen. From a side view, DPC-2803 displays the shortened nose of anthropoids, the dished or sloping appearance of the face (note the very sloping appearance of the face in the orangutan), and large upper canine sockets. *Aegyptopithecus* appears to have been sexually dimorphic (large males with smaller females) as indicated by the large and small sizes of the canine teeth collected at the Fayum. Presence of sexual dimorphism in this genus implies the presence of social organization involving male competition. Like Old World monkeys and apes, the dental formula is 2-1-2-3. The molars have low rounded cusps indicative of fruit eating and a Y-5 cusp pattern is present on the lower molars.

All indications from the skeleton of *Aegyptopithecus* suggest that it was a small primate (weighing about 13 pounds), arboreal, and moved using quadrupedalism and climbing. In contrast to living apes, no knuckle-walking or arm-swinging adaptations have been observed in *Aegyptopithecus*. Some researchers believe *Aegyptopithecus* is the ancestor of all later hominoids (for example, *Proconsul* and chimpanzees), and others believe that *Aegyptopithecus* is a more primitive catarrhine and is ancestral to both Old World monkeys and to all later hominoids. *Aegyptopithecus* is believed to have lived in a lowland rainforest along numerous rivers and lakes with seasonal heavy rains. This ecological context is in sharp contrast to the Sahara desert of today.

References: Simons, 1967, 1972, 1987, 1995; Radinsky, 1973; Kay and Simons, 1980; Kay, Fleagle, and Simons, 1981; Bown, Kraus, Wing, and others, 1982; Fleagle and Simons, 1982a, 1982b; Fleagle 1983; Fleagle and Kay, 1983; Fleagle, Bown, Obradovich, and Simons, 1986; Olson and Rasmussen, 1986; Gebo and Simons, 1987; Gebo, 1989, 1993; Conroy, 1990; Kappelman, 1992.

DPC–1028 (mandible) DPC-2803 (cranium)

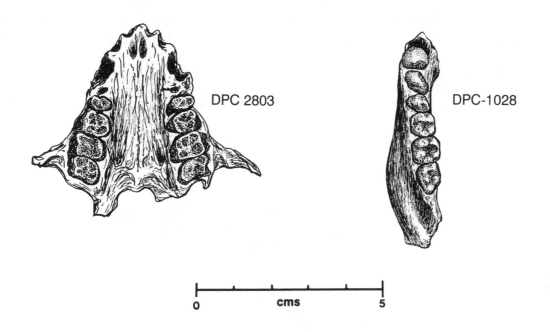

DPC 2803

DPC-1028

0   cms   5

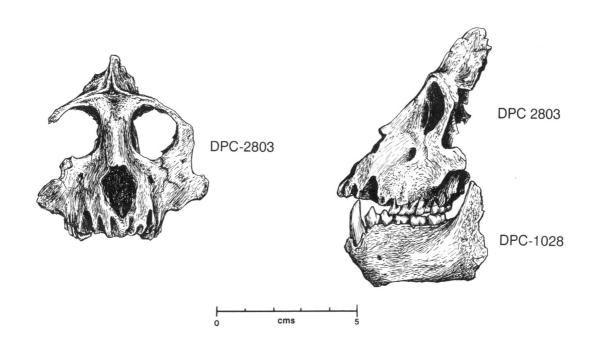

DPC-2803

DPC 2803

DPC-1028

0   cms   5

KNM-RU 7290

Specimen: **KNM-RU 7290**
Geographic location: Rusinga Island,
  Lake Victoria, Kenya

Taxonomic affiliation: *Proconsul heseloni*
Dating: Early Miocene (18 myBP)

General Description: Mary Leakey discovered the cranial, facial, and postcranial remains of this partial skeleton of a subadult individual in 1948. Other fragments belonging to this skeleton were discovered in 1951. Previously unidentified cranial fragments found in the collections of the National Museums of Kenya have allowed a fuller reconstruction of the partially crushed skull (Walker, Falk, Smith, and Pickford, 1983). Originally called *Proconsul africanus*, Walker and coworkers (1993) have revised the taxonomy of the dental specimens from the East African sites of Koru, Songhor, Rusinga Island, and Mfangano Island. They conclude that the *Proconsul* fossils from Koru and Songhor belong to a single species, *Proconsul africanus*, and the fossils from Rusinga and Mfangano Islands belong to two species, a larger *Proconsul nyanzae* and a smaller *Proconsul heseloni*. Thus, all of the small fossils from Rusinga Island, including the skull and limb bones attributed to KNM-RU 7290, formally belong to the newly-named species, *Proconsul heseloni*.

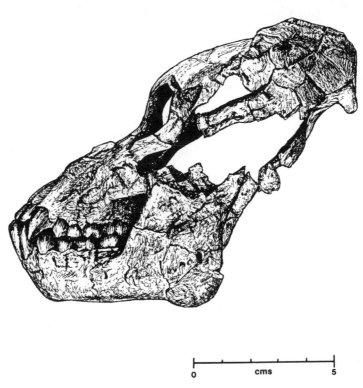

Note the very small browridges, the relatively projecting snout, and the prominent canine alveoli. This cranium also possesses a tall and narrow nasal aperture rather than a broad and rounded one as in living apes. *Proconsul heseloni* possesses ape-like teeth with a 2-1-2-3 dental formula, a Y-5 molar cusp pattern, low rounded cusps on the molars, sexually dimorphic canines, and sectorial lower premolars. Brain size and body size (33–44 pounds) are much larger in *Proconsul heseloni* than in *Aegyptopithecus*.

Fieldwork in the 1980s produced a wealth of new postcranial materials from this locality. The postcranium of *Proconsul heseloni* is very primitive compared to living apes. The shoulder, elbow, and wrist lack the specializations which allow for the complex rotatory movements of these joints (*Proconsul heseloni* was incapable of brachiating). This is further supported by the more equal limb lengths in this taxon, and thus *Proconsul heseloni* lacks the long arms characteristic of living apes. *Proconsul heseloni* is, however, ape-like with its large fibula, gibbon-like foot anatomy, and no tail (Ward and others, 1991). Although still under debate, the phylogenetic position of *Proconsul heseloni* is likely to be before the separation of the gibbon lineage. Therefore, it is not closely allied with the great apes as proposed previously.

References: Le Gros Clark and L. Leakey, 1951; Napier and Davis, 1959; Davis and Napier, 1963; Simons, 1972; L. Leakey, 1974; Walker, Falk, Smith, and Pickford, 1983; Falk, 1983c; Fleagle, 1983, 1988; Rose, 1983; Walker and Pickford, 1983; Andrews, 1985; Langdon, 1986; Kelley, 1986; Beard, Teaford, and Walker, 1986; Harrison, 1987; Gebo, Beard, Teaford, and others, 1988; Walker, Teaford and R. Leakey, 1986; Ruff, 1988; Teaford, Beard, R. Leakey, and Walker, 1988; Ruff, Walker, and Teaford, 1989; Walker and Teaford, 1989; Ruff, Walker, and Teaford, 1989; Conroy, 1990; Ward, Walker, and Teaford, 1991; Ward, Walker, Teaford, and Odhiambo, 1993; Ward, 1993; Rose, 1993; Rose, Leakey, Leakey, and Walker, 1992; Walker, Teaford, Martin, and Andrews, 1993; Begun, 1994; Begun, Teaford, and Walker, 1994; Retallack, Bestland, and Dugas, 1995; Benefit and McCrossin, 1995; Gebo, 1996.

KNM-WK 16999

Specimen: **KNM-WK 16999**
Geographic location: Kalodirr, Kenya

Taxonomic affiliation: *Afropithecus turkanensis*
Dating: Middle Miocene (16-18 myBP)

General Description: This facial skeleton and associated mandibular and postcranial remains were found in a locality near Lake Turkana in northern Kenya and described by Richard and Meave Leakey in 1986. These remains are from a large hominoid which possessed a relatively long snout, long and steep nasal bones, a wide interorbital distance, thin browridges, and a sagittal crest with temporal lines which merge together on the frontal just above the browridges. This specimen possesses incisors which are procumbent with large and broad central incisors and large tusk-like canines. The palate is shallow, the tooth rows are parallel, and the jaw is deep and robust. In a number of respects, the face of *Afropithecus* resembles that of *Aegyptopithecus*. Other aspects of its anatomy show affinities to other Miocene hominoids (e.g., "*Heliopithecus*" and *Kenyapithecus*).

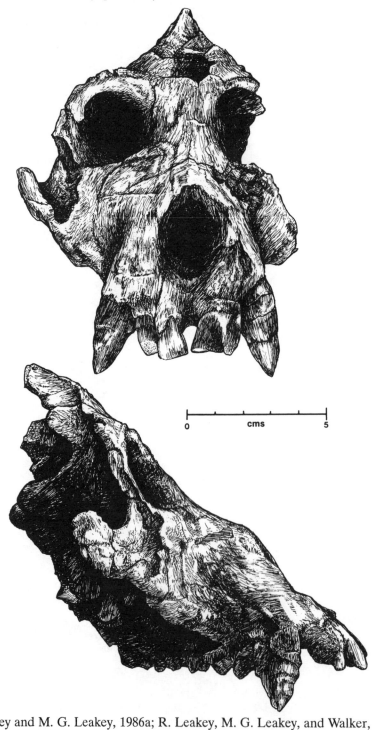

References: R. Leakey and M. G. Leakey, 1986a; R. Leakey, M. G. Leakey, and Walker, 1988b; Andrews, 1992; Begun, 1994; Benefit and McCrossin, 1995.

UM.P. 62-11

Specimen: **UM.P. 62-11**
Geographic location: Moroto II, Karamoja
  District, Uganda

Taxonomic affiliation: *Morotopithecus bishopi*
Dating: Early Miocene (>20.6 myBP)

General description: This well preserved maxilla and upper dentition were collected in a series of expeditions (1961–1963) led by W. W. Bishop and F. Whyte (Uganda Museum and University of Glasgow) and D. Allbrook (Makerere University College). This maxilla represents a very large hominoid which was initially allocated to the larger species of *Proconsul* (*P. major*) and later to *Afropithecus sp.* (R. Leakey and others, 1988b). Reassessment of the fossil assemblage from Moroto and other east African Miocene hominoids, however, indicates the presence of a new genus and species, *Morotopithecus bishopi* (Gebo and others, 1997). UM.P. 62-11 shows impressively large canines that are suggestive of a male hominoid and a number of other dental features common to Miocene hominoids (e.g., 2-1-2-3 dental formula, bunodont cusps, parallel tooth rows, diastema, narrower central incisors, beaded cingulum). Compare this specimen with the dentition of modern pongids. Note the more blade-like canines, the larger central incisors, the broader premolars, and the sharper molar cusps found in gorillas compared to UM.P. 62-11. Postcrania recently recovered from Moroto by Gebo and coworkers confirms the great robusticity of this early hominoid. Based on the study of postcrania, it appears that the lower back of this hominoid was more erect (orthograde)—like that of living apes and humans, and unlike that of the more monkey-like body plan of *Proconsul*. Similarly, the shoulder region is more like living apes and humans. The overall morphology of the shoulder implies enhanced arm mobility and is associated with apes that brachiate or arm-hang. *Morotopithecus bishopi* represents the oldest hominoid sharing derived body characteristics with living apes and humans.

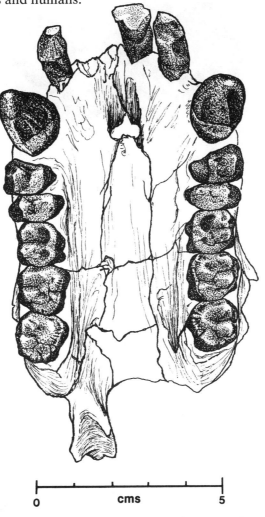

0   cms   5

References: Allbrook and Bishop, 1963; Bishop, 1964; Bishop, Miller, and Fitch, 1969; Simons, 1972; R. Leakey, M. G. Leakey, and Walker, 1988b; Fleagle, 1988; Andrews, 1992; Ward, 1993; Saunders and Bodenbender, 1994; Begun, 1994; Benefit and McCrossin, 1995; MacLatchy and Bossert, 1996; Pilbeam, 1996; Gebo, MacLatchy, Kityo, Deino, Kingston, and Pilbeam, 1997.

KNM-WK 16950A

Specimen: **KNM-WK 16950A**
Geographic location: Kalodirr, Kenya

Taxonomic affiliation: *Turkanapithecus kalakolensis*
Dating: Middle Miocene (16-18 myBP)

General description: This specimen is a fairly complete cranium which is associated with mandibular and postcranial elements. It was briefly described by Richard and Meave Leakey in 1986 and more fully described by R. Leakey, M. G. Leakey, and Walker (1988a). It is from the same locality as *Afropithecus*, but represents a much smaller hominoid that is similar in size to *Proconsul*. This cranium possesses the following features: a broad snout, small orbits, a shallow palate, a wide interorbital region, thin brow-ridges, and it lacks a sagittal crest. The tooth rows are parallel. The mandible is shallow with no inferior transverse torus and a low mental foramen. The canine is large, the enamel is thin, and the molars exhibit evidence of even tooth wear. This specimen possesses a puncture mark from a carnivore just above the right canine. *Turkanapithecus* is an early hominoid, but it lacks clear taxonomic relationship with the other known hominoids of the Miocene.

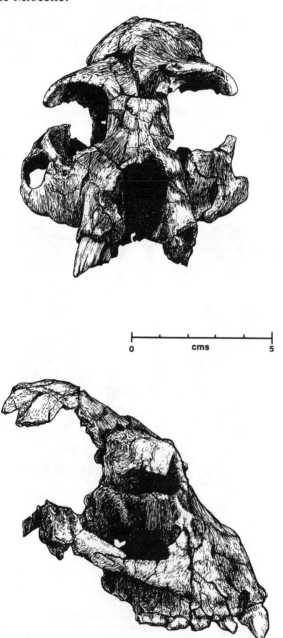

References: R. Leakey and M. G. Leakey, 1986b; R. Leakey, M. G. Leakey, and Walker, 1988a; Conroy, 1990; Begun, 1994; Benefit and McCrossin, 1995.

Specimen: **M.N.H.N.P.-A.C. 36**
Geographic location: Saint Gaudens, France

Taxonomic affiliation: *Dryopithecus fontani*
Dating: Middle to late Miocene (14 myBP)

General description: The type specimen of *Dryopithecus fontani,* a mandible, was first discovered near the village of Saint Gaudens, France, in 1856 by M. Fontan. It was later described by E. Lartet. Other specimens of *Dryopithecus* have been found throughout Europe. Dentally, these specimens show large canines with a honing facet on the lower premolar, similar wear on the molars (reflecting a rapid molar eruption sequence), thin enamel, a 2-1-2-3 dental formula, and low rounded cusps adapted for fruit eating. The incisors are small, indicating less stripping and husking adaptations for eating fruits with their front teeth. This feature contrasts with very broad central incisors that are present in living apes. The few postcranial specimens that have been found are more ape-like for *Dryopithecus* than for *Proconsul. Dryopithecus* possesses a reduced olecranon process and a deep humeral trochlea. Both features are similar to living apes and are functionally related in allowing the elbow joint to fully extend the forearm (an adaptation for arm suspension). This partial skeleton from Can Llobateres, Spain, shows long arms and fingers for *Dryopithecus laietanus,* features that are very similar to the living apes. Thus, dryopithecines are more similar to living apes than are the early Miocene hominoids.

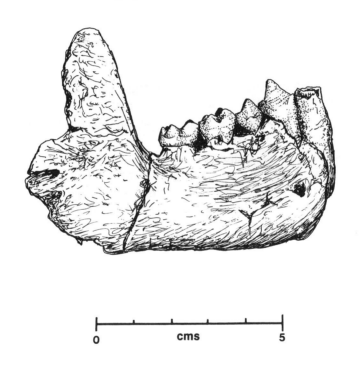

References: Lartet, 1856; Simons and Pilbeam, 1965; Simons, 1972; Szalay and Delson, 1979; Fleagle, 1983, 1988; Morbeck, 1983; Kelley and Pilbeam, 1986; Andrews and Martin, 1987; Begun, Moyà Solà, and Köhler, 1990; Conroy, 1990; Begun, 1993, 1994; Moyà Solà and Köhler, 1993; Begun, 1994; Benefit and McCrossin, 1995; Kordos and Begun, 1997.

Specimen: **IGF 11778**

Geographic location: Monte Bamboli, Tuscany, Italy

Taxonomic affiliation: *Oreopithecus bambolii*

Dating: Late Miocene (8 myBP)

General description: *Oreopithecus* was discovered in coal mines from sites in northern Italy and was first described by Gervais in 1872. Other remains now include cranial and postcranial elements, but unfortunately, the best known skeleton is crushed (see following page), thus limiting its usefulness for interpretation. *Oreopithecus* has a 2-1-2-3 dental formula, dimorphic canines, premolar honing, parallel tooth rows like other hominoids, but possesses unusually large and round upper central incisors, well-developed shearing crests (adaptations for leaf-eating), and a centroconid on the lower molars. Postcranially, *Oreopithecus* lacks a tail, and possesses very long arms with curved fingers and toes, a broad thorax with a short trunk, and a modern ape-like shoulder and elbow joints adapted for suspension. All of these postcranial features are similar to extant apes and differ especially from early Miocene hominoids.

References: Gervais, 1872; Hürzeler, 1958, 1960; Straus, 1963; Simons, 1972; Szalay and Berzi, 1973; Szalay and Delson, 1979; Delson, 1986a; Azzaroli, Boccaletti, Delson, Moratti, and Torre, 1986; Szalay and Langdon, 1986; Harrison, 1986; Jungers, 1987; Sarmiento, 1987; Fleagle, 1988; Benefit and McCrossin, 1995; Conroy, 1990; Rook, Harrison, and Engesser, 1996; Harrison and Rook, 1997.

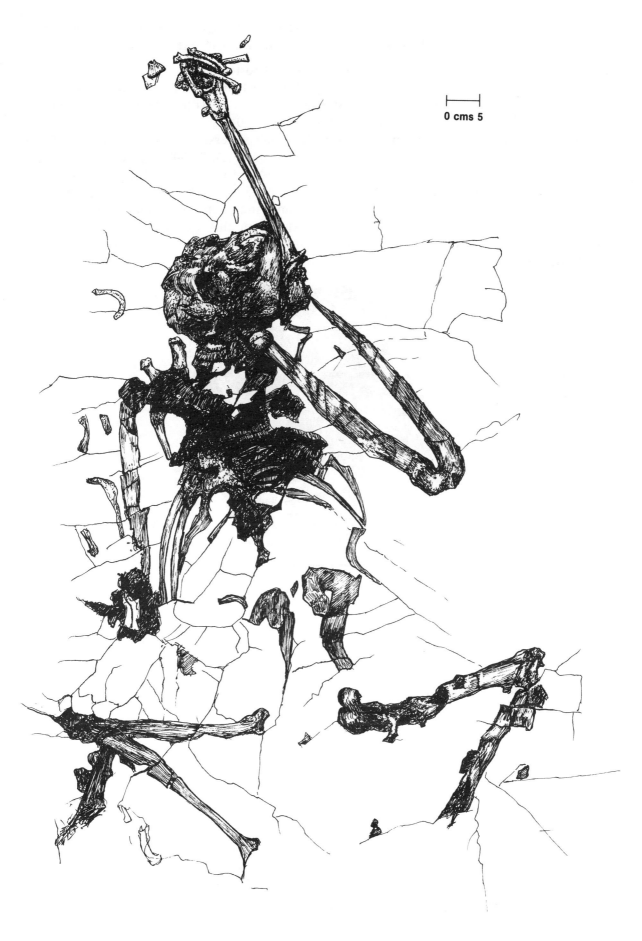

0 cms 5

AS95-500

Specimen: **AS95-500**
Geographic location: Yassiören, Turkey

Taxonomic affiliation: *Ankarapithecus meteai*
Dating: Late Miocene (9.8 myBP)

General description: In 1995, a partial skull, jaw and several postcranial elements were recovered from Locality 12 in the Sinap Formation in central Turkey. This new late Miocene material, along with material collected over the past 40 years, has been systematically revised within the taxon, *Ankarapithecus*, rather than *Sivapithecus*. Due to the completeness of AS95-500, several cranial characteristics can be analyzed to better understand great ape evolution. AS95-500 possesses an interesting array of features that are unlike any single fossil or living great ape. For example, AS95-500 has a very large maxillary central incisor, relative to its lateral incisor, and a relatively narrow interorbital region. Both of these features are like that of orangutans (see the facial view of the orangutan shown earlier for a comparison). On the other hand, AS95-500 possesses thick molar enamel and a very robust jaw, anatomical features that resemble other fossil apes, such as *Sivapithecus*. At present, this complex array of cranial features appears to link *Ankarapithecus* as a stem member of the great ape and human lineages rather than any specific association with orangutans, gorillas, or chimpanzees.

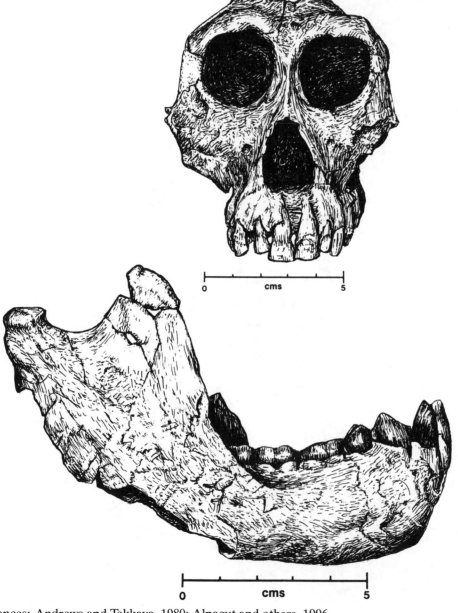

References: Andrews and Tekkaya, 1980; Alpagut and others, 1996.

IVPP PA580

Specimen: **IVPP PA580**
Geographic location: Lufeng County,
  Yunnan Province, People's Republic of China

Taxonomic affiliation: *Lufengpithecus lufengensis*
Dating: Late Miocene (8 myBP)

General description: The site from which this specimen was recovered has produced the greatest number of Miocene hominoids in Asia. Over a thousand teeth, a number of crania and mandibles, and some postcranial elements have been found at this locality by Chinese paleontologists from a series of expeditions beginning in 1975. This mandible was recovered in 1976 by a paleontological team from the Institute of Vertebrate Paleontology and Paleoanthropology (Chinese Academy of Sciences) and the Yunnan Provincial Museum. Although crushed and slightly deformed, this specimen is quite complete and is missing only the central incisors and is thought to represent a female of this sexually dimorphic species (Kelley and Etler, 1989). This specimen possesses relatively small incisors, slender canines like *Dryopithecus*, occlusal crenelations on its molars, and no wear gradient. IVPP PA580 appears more similar to the European dryopithecines (Wu, 1987; Kelley and Etler, 1989) than to *Ramapithecus* or *Sivapithecus* as previously believed by Xu, Lu, Pan, Qi, Zhang, and Zheng (1978), Kay (1982), Kay and Simons (1983), and Wu and Oxnard (1983).

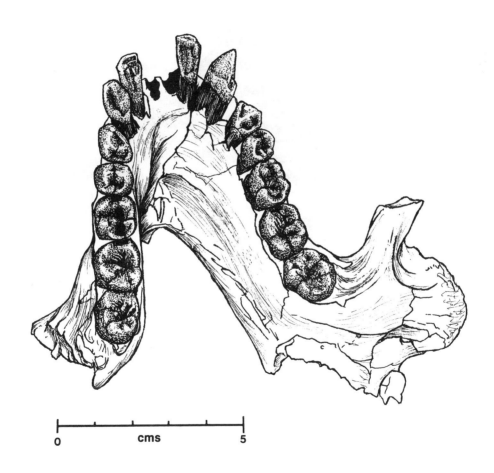

0   cms   5

References: Xu Qinghua, Lu Qingwu, Pan Yuerong, Qi Guoqin, Zhen Xingyong, and Zheng Liang, 1978; Xu Qinghua and Lu Qingwu, 1979; Institute of Vertebrate Paleontology (Chinese Academy of Sciences), 1980; Wolpoff, 1982, 1998; Kay, 1982; Kay and Simons, 1983; Wu and Oxnard, 1983; Etler, 1984; Wu Rukang and Xu Qinghua, 1985; Kelley and Pilbeam, 1986; Wu, 1987; Fleagle, 1988; Kelley and Etler, 1989; Conroy, 1990; Wu and Poirier, 1995; Benefit and McCrossin, 1995.

Specimen: **IVPP PA644**

Geographic location: Lufeng County,
 Yunnan Province, People's Republic of China

Taxonomic affiliation: *Lufengpithecus lufengensis*

Dating: Late Miocene (8 myBP)

General description: This badly crushed specimen—including a complete cranium and dentition—was recovered in 1978 by a joint Institute of Vertebrate Paleontology and Paleoanthropology—Yunnan Provincial Museum expedition. Although the distortion of the face has made study difficult, this specimen exhibits an interesting array of characteristics which are similar to fossils found in northern Kenya at Buluk (Walker and R. Leakey, 1984), from the Potwar Plateau of Pakistan (Pilbeam, 1982), and from Turkey (Andrews and Tekkaya, 1980; McHenry, Andrews, and Corruccini, 1980). Numerous workers have noted similarities with *Sivapithecus*, and Xu and Lu (1980) have named this specimen *Sivapithecus yunnanensis*. Others believe it belongs to *Sivapithecus indicus* (Kay and Simons, 1983). More recently, work has shown this specimen to be more primitive than *Sivapithecus* in craniodental morphology, and this specimen has been renamed *Lufengpithecus lufengensis* along with the smaller (presumably female) dental remains from Lufeng (Wu, 1987; Kelley and Etler, 1989). IVPP PA644 possesses large canines, broad central incisors, deep and flaring zygomatics, prominent temporal lines which meet to form a sagittal crest, a broad interorbital distance, and moderately thin browridges. Postorbital constriction is also evident. Compared to GSP 15000 (*Sivapithecus indicus*), IVPP PA644 possesses a short and broader face with more robust lateral orbital margins, a wider interorbital distance, slender canines, and narrower central incisors (Kelley and Pilbeam, 1986; Kelley and Etler, 1989).

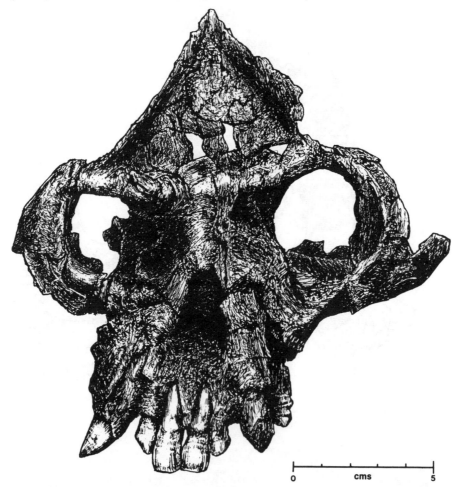

0   cms   5

References: Xu Qinghua and Lu Qingwu, 1979, 1980; Institute of Vertebrate Paleontology and Paleoanthropology (Chinese Academy of Sciences), 1980; Lu Qingwu, Xu Qinghua, and Zheng Liang, 1981; Wu Rukang, Han Defen, Xu Qinghua, Lu Qingwu, Pan Yuerong, Zhang Xingyong, Zheng Liang, and Xiao Minghua, 1981; Kay, 1982; Wolpoff, 1982; Pilbeam, 1982; Wu Rukang, Xu Qinghu, and Lu Qingwu, 1983; Kay and Simons, 1983; Hammond, 1983; Etler, 1984; Rensberger, 1984; Wu Rukang and Xu Qinghua, 1985; Kelley and Pilbeam, 1986; Wu Rukang, 1987; Kelley and Etler, 1989; Conroy, 1990; Wu and Poirier, 1995; Benefit and McCrossin, 1995.

GSP 15000

Specimen: **GSP 15000**
Geographic location: Potwar Plateau, Pakistan

Taxonomic affiliation: *Sivapithecus indicus*
Dating: Late Miocene (8 myBP)

General description: This skull was recovered during the 1979–1980 field season by D. Pilbeam and S. M. Ibrahim Shah (Geological Survey of Pakistan-Yale Peabody Museum Siwalik research project). It consists of the left side of the face with a nearly complete mandible and a complete dentition. The overall structure of the face includes a combination of delicate, thin bone coupled with large jaws and teeth. The face is very orang-like in a number of features: dished facial profile with upraised incisor alveoli, subnasal morphology, oblong orbits, narrow interorbital distance, lack of browridge development, and large size of the central incisors. *Sivapithecus* possesses a 2-1-2-3 dental formula, thick enamel, and loss of the cingulum around the base of the molars. The cheek teeth are closely packed together and interstitial wear is evident. A wear gradient is present indicating greater time spans between molar eruptions. The dentition reflects a diet of hard-fibrous foods.

The postcranial evidence for *Sivapithecus* shows it to be a quadrupedal and climbing oriented ape. It possesses a mobile hip joint, a grasping big toe, curved toe bones (which are not especially long like an orang's), and a modern ape-like elbow. Its wrist and foot anatomy suggest a semiterrestrial primate. The arm bones (humeri) of *Sivapithecus* are more controversial. These bones are robust and medially curved like terrestrial primates, rather than straight like an ape's. The arm bones also lack a proximal end (in the region of the shoulder joint), thus we are not sure how mobile (and ape-like) the shoulder was. Initially, *Sivapithecus* was thought to be closely aligned with the hominid lineage due to their similarities in dentition, but the facial evidence has altered this interpretation and placed *Sivapithecus* closer to orangutans. Recent study of the postcranial evidence has shown that this fossil hominoid is not especially orang-like below its neck. Thus, the systematic position of *Sivapithecus* remains unclear.

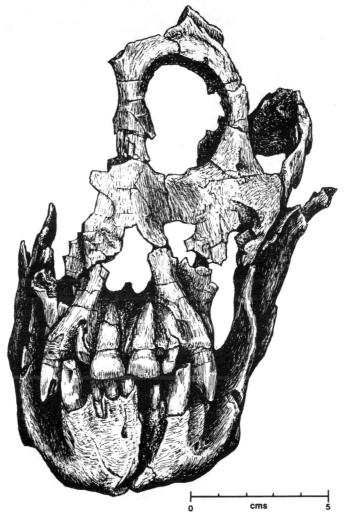

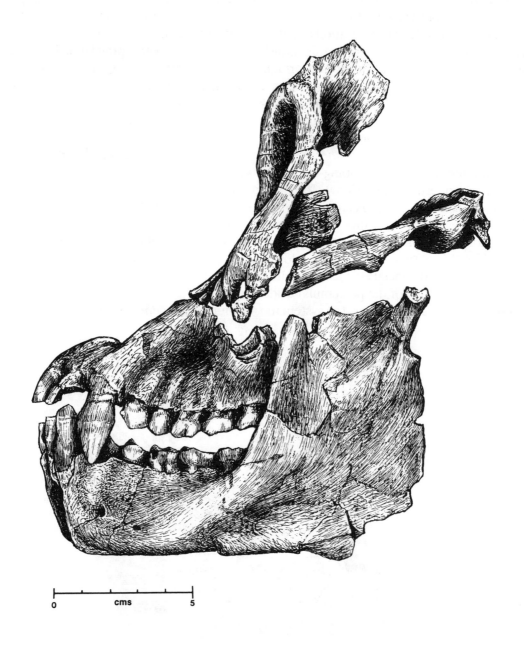

GSP 15000

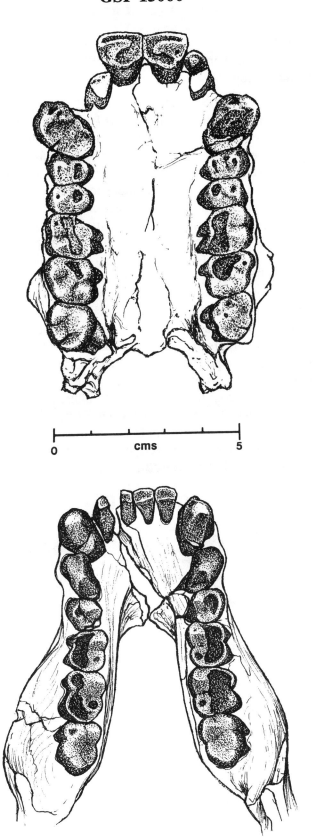

0   cms   5

References: Pilbeam, 1982, 1984b, 1989; Andrews, 1982, 1983; Andrews and Cronin, 1982; Preuss, 1982, Wolpoff 1982, 1983c; Ward and Kimbel, 1983; Kay and Simons, 1983; Ward and Pilbeam, 1983, Ward and Brown, 1986; Rose, 1983, 1986, 1989, 1993; Rensberger, 1984; Schwartz, 1984a, 1984b; Badgley, Kelley, Pilbeam, and Ward, 1984; Langdon, 1986; Kelley and Pilbeam, 1986; Martin, 1986; Gebo, 1989, 1996; Conroy, 1990; Pilbeam, Rose, Barry, and Ibrahim Shah, 1990; Spoor, Sondaar, and Hussain, 1991; Benefit and McCrossin, 1995; McCollum and Ward, 1997.

Mandibles II and III

Specimen: **Mandibles II and III**
Geographic location: Tahsin and Liuchang
  Districts of Kwangsi Province, People's
  Republic of China

Taxonomic affiliation: *Gigantopithecus blacki*
Dating: Pleistocene (0.5–1.0 myBP)

General description: Initially, the teeth of *Gigantopithecus blacki* were found in drugstores in China by G.H.R. von Koenigswald in 1935 where they were sold as "dragon teeth" for medicine. In the 1950s and 1960s, Chinese scientists recovered several jaws of this large ape from southern China. In 1968, a paleontological team from Yale University working in India discovered a smaller, but very similar, jaw from late Miocene deposits (approximately 8 myBP). This specimen was named *Gigantopithecus bilaspurensis* (Simons and Ettel, 1970). Kay and Simons (1983) subsequently placed this specimen within the taxon *Sivapithecus giganteus*, denoting its close similarity to *Sivapithecus*. *Gigantopithecus blacki* is only known from jaws and teeth which are of immense size. This Pleistocene ape, a contemporary of the hominid *Homo erectus*, is thought to have weighed twice the size of an adult male gorilla and stood between 8 and 9 feet tall when upright. No undisputed postcranial remains are known for *Gigantopithecus*. Therefore, it is difficult to confirm these large size estimates or reconstruct locomotor capabilities.

  The jaws of *Gigantopithecus* are very thick and deep. There is no diastema, and thus the teeth are closely packed together. The tooth formula is 2-1-2-3. The incisors are relatively small and the canines wear flat across their tops. The premolars are molarized and thus look like the very large and flat molars along the tooth row. The lower molars possess a Y-5 pattern, thick enamel, and there is a wear gradient from M1 to M3 indicating a delayed eruption sequence. Functionally, the robust jaw and the heavy emphasis on back teeth rather than front teeth suggest a dramatic shift in diet compared to other fruit-eating apes. *Gigantopithecus* is a specialist for heavy chewing. Some believe its diet consisted of hard nuts and fruits. Others have argued that it chewed bamboo in the same fashion as panda bears. This dietary regime, jaw size, and dental adaptations are believed to be a hyper-development along the dietary trend and dental adaptations first observed in *Sivapithecus*.

G. blacki II

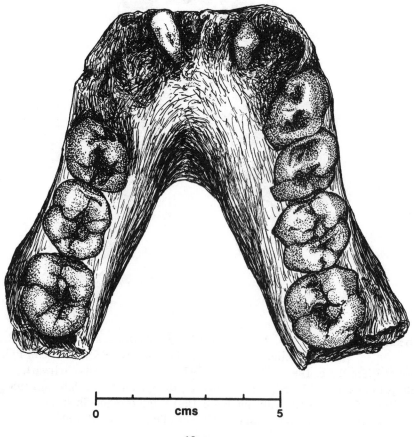

0   cms   5

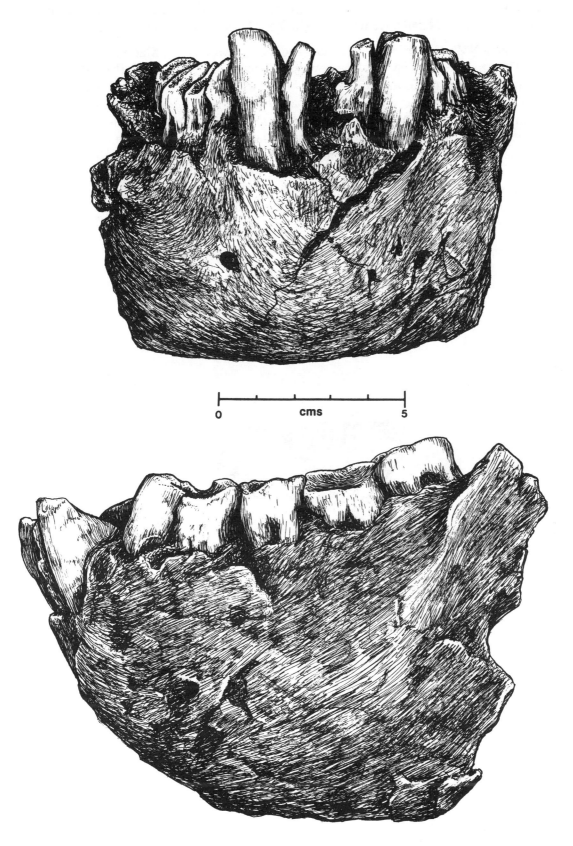

0   cms   5

References: Weidenreich, 1945; von Koenigswald, 1952; Simons and Chopra, 1969; Pilbeam, 1970; Simons and Ettel, 1970; Simons, 1972; White, 1975; Frayer, 1973; Gelvin, 1980; Yinyun, 1982; Kay and Simons, 1983; Zhang Yinyun, 1985; Fleagle, 1988; James, 1989; Wu and Poirier, 1995; Benefit and McCrossin, 1995.

KNM-MB 20573

Specimen: **KNM-MB 20573**
Geographic location: Maboko Island, Kenya

Taxonomic affiliation: *Kenyapithecus africanus*
Dating: Middle Miocene (14.7–16 myBP)

General description: KNM-MB 20573 is a juvenile mandible recovered from Maboko Island (Lake Victoria), Kenya. This specimen, along with controversial specimens recovered in the 1960s from Fort Ternan, Kenya, represent the genus *Kenyapithecus*. This genus has been difficult to diagnose taxonomically relative to *Sivapithecus*, but several new studies and discoveries have helped to clarify the distinctiveness of this ancient ape. KNM-MB 20573 is distinctive in its possession of a robust symphysis with a procumbent lower incisor. These features, along with its large canines, suggest an important feeding component for the anterior dentition. Foods such as hard nuts or tough-skinned fruits, where incisor preparation is essential and needs to be forceful, may have made up a large part of the diet of *Kenyapithecus*. The molars also support this feeding interpretation. The molar morphology of *Kenyapithecus* is low-crowned, crenulated on their occlusal surfaces, they lack cingula, and they are thick-enameled with a distinctive wear gradient.

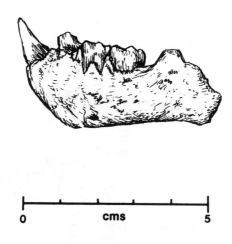

0   cms   5

References: Andrews, 1971; Andrews and Walker, 1976; Kelley and Pilbeam, 1986; Martin, 1986; Ward and Brown, 1986; McCrossin and Benefit, 1993.

BER I, 1'91

Specimen: **BER I, 1'91**
Geographic location: Berg Aukas, Namibia

Taxonomic affiliation: *Otavipithecus namibiensis*
Dating: Middle Miocene (13 myBP)

General description: In 1991, Glenn Conroy and colleagues found an ape jaw in cave breccias in southern Africa, the first fossil hominoid below the equator. The jaw comes from a 10-year-old individual that weighed between 30 and 44 pounds. The molars are puffy and bunodont in outline and small in size relative to the deep and robust mandible. Little differential wear can be found on the molars, indicating a similar maturational development to that of chimpanzees. Further, the molars lack cingula and the first molar is much smaller than the other two. Interestingly, the molars of *Otavipithecus* are thin-enameled, like that of African apes. The jaw also possesses a narrow incisor region. All of the dental and mandibular features suggest a diet of non-abrasive foods such as leaves, berries, seeds, buds, and flowers.

The phyletic position of this specimen is controversial. It shares several dental features with other Miocene hominoids, such as *Dryopithecus* and *Kenyapithecus*, but it is also unlike other middle Miocene hominoids in a variety of features. One interesting evolutionary interpretation suggests a close evolutionary position with that of African apes, but the phyletic position of this fossil is unresolved.

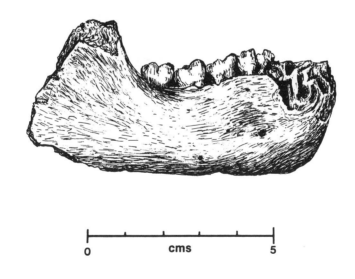

0   cms   5

References: Conroy, Pickford, Senut, Van Couvering, and Mein, 1992; Andrews, 1992; Begun, 1994; Conroy, 1994; Conroy, Lichtman, and Martin, 1995.

Australopithecus

0   cms   5

STS 5

3. AUSTRALOPITHECUS

Pliocene-Pleistocene Hominids

The first fossils that are indisputably hominid are from the Pliocene epoch dating to well before 4 million years ago. The earliest evidence for these hominids has been recovered by paleontological teams in the Middle Awash Valley, Ethiopia, at Aramis, Maka, and Belohdelie (White, 1984; Clark and others, 1984: Hill and Ward, 1988; White and others, 1993, 1994, 1995) as well as at several other east African localities (Hill and Ward, 1988; M. Leakey and others, 1995; Wolpoff, 1998). Paleoanthropologists recognize three species of early hominids—*Ardipithecus ramidus, Australopithecus anamensis, Australopithecus afarensis*—which date to 4.4 myBP at Aramis, Ethiopia, between 3.9 myBP and 4.2 myBP at Kanapoi and Allia Bay, Kenya, and 3.75 myBP at Laetoli, Tanzania, respectively. *Ardipithecus*, the earliest hominid, is primitive, resembling pongids in several details of the cranial, dental, and postcranial anatomy. *A. afarensis* is also well represented at Hadar, Ethiopia, from deposits that date to between 3.0 and 3.4 myBP (Kimbel and others, 1994) and it has been documented from a single specimen recovered from Koobi Fora on the eastern side of Lake Turkana, Kenya (Kimbel, 1988). Analysis of footprints at Laetoli and the locomotor skeleton—especially the hip, knee, and ankle—shows clear evidence for habitual bipedality in *A. afarensis*. The best evidence suggests that *Ardipithecus* gave rise to *A. afarensis*, which served as the ancestor for the evolution of later australopithecines and *Homo*. Analysis of the geology and fossil plants and animals associated with *Ardipithecus* suggests that the earliest hominids had their origin in a wooded setting (WoldeGabriel and others, 1994), challenging a long held assumption that the earliest hominids arose in an open, grasslands environment.

Later australopithecines include *Australopithecus africanus, Australopithecus aethiopicus, Australopithecus robustus*, and *Australopithecus boisei*. For reasons that remain unclear, *A. boisei* appears to have gone extinct between 1.2 and 0.7 million years ago (Klein, 1988). Taken together, there appears to be at least three lineages of hominids. Two lineages of australopithecines show an ever increasing focus on heavy use of the masticatory apparatus. This increasing specialization is illustrated by the increasing size and changing orientation of the chewing musculature and supporting bony architecture. In particular, we see a forward shift of the chewing muscles and bones of the cranium that support these muscles. In keeping with these changes, there is a marked expansion of the postcanine dentition. The greatest expression of this adaptive complex is seen in *A. aethiopicus* and *A. boisei*. The early date for *A. aethiopicus* (2.5 myBP) demonstrates that very large cheek teeth and other adaptations to heavy chewing occurred early in the evolution of the australopithecines. There seems to have been little or no increase in brain size in comparison of early and late australopithecines, reflecting perhaps little change in intelligence. Study of *A. robustus* hand bones, however, suggests that they may have been both tool-users and tool-producers (Susman, 1988, 1994).

The site locations for the australopithecines that are presented here include the following materials:
1. Laetoli, Tanzania: L.H. 4
2. Hadar, Ethiopia: A.L. 288-1, A.L. 200-1a, A.L. 400-1a, Hadar cranial
   reconstruction, A.L. 444-2
3. Taung, South Africa
4. Sterkfontein, South Africa: STS 5, STS 14, STS 52a, STS 52b, STS 71
5. Swartkrans, South Africa: SK 46, SK 23, SK 48
6. Olduvai Gorge, South Africa: O.H. 5
7. Peninj, Tanzania
8. East Turkana, Kenya: KNM-ER 406, KNM-ER 732
9. West Turkana, Kenya: KNM-WT 17000

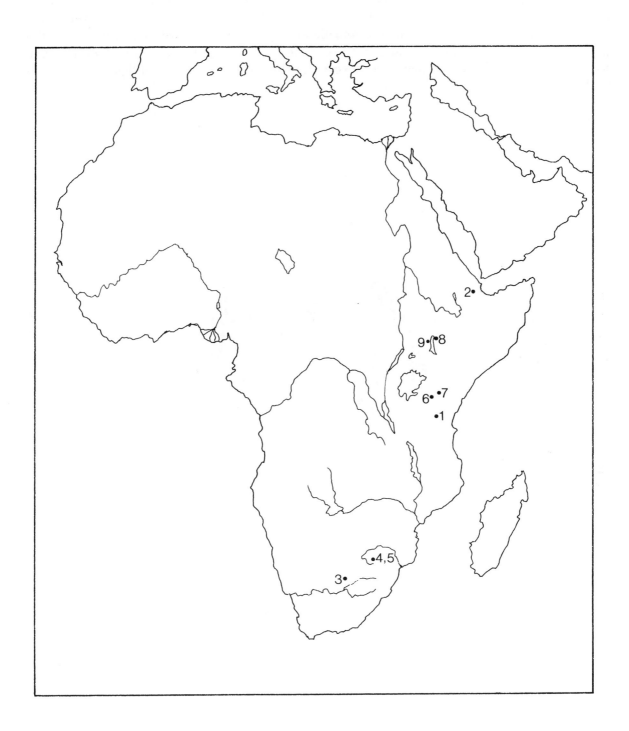

Specimen: **L.H. 4**

Geographic location: Laetoli, Tanzania

Taxonomic affiliation: *Australopithecus afarensis*

Dating: Middle Pliocene (3.6–3.8 myBP)

General description: This mandible, along with part of the dentition, was discovered by M. Muluila working with Mary Leakey in 1974. It is one individual of a sample of fossils that are representative of some of the earliest known hominids in Africa. This individual and others from Laetoli were described by T. White. The primitive features of these remains are especially impressive. Although not functionally analogous to the diastema of apes, the gap between the canine and premolar is primitive. The first premolar is also primitive, more so than other species of hominid. The narrow, oval shape of this tooth and the dominant single cusp are pongid-like.

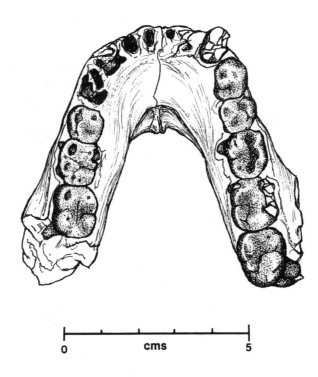

0   cms   5

References: M. D. Leakey, Hay, Curtis, Drake, Jackes, and White, 1976; White, 1977, 1980; Johanson, White, and Coppens, 1978; Johanson and White, 1979; Johanson and Edoy, 1981; White, Johanson, and Kimbel, 1981; M. D. Leakey, 1984; Harris, 1985; M. D. Leakey and Harris, 1987; Tuttle, 1988.

A.L. 288-1

Specimen: **A. L. 288-1 ("Lucy")**
Geographic location: Hadar, Ethiopia

Taxonomic affiliation: *Australopithecus afarensis*
Dating: Middle Pliocene (3.0 myBP)

General description: Discovered in 1974 by D. Johanson and T. Gray, this 40% complete fossil hominid skeleton represents one of the most complete individual remains of an australopithecine. The individual is a very small female, weighing about 60 pounds and standing not much greater than three and a half feet in stature. In addition to being short, this individual possesses a suite of features that are very different from that of living humans. The bones are extraordinarily thick, reflecting a demanding, strenuous lifestyle. The relative elongation of the bodies of the thoracic vertebra reflect heavy use of the back, such as climbing, lifting, and carrying. The relative lengths of the arms and legs are different from those of modern humans. William Jungers contends that these differences reflect a hominid that had attained arm proportions similar to modern humans, but the legs were much shorter (although see Wolpoff, 1983a, 1983b). This implies that this hominid had a short stride length, thus requiring more steps to move the same distance relative to a modern human.

The curved hand and foot phalanges are reminiscent of those of living apes. Stern and Susman (1983) have remarked that the curvature of the Hadar proximal hand phalanges—like chimpanzees—is very pronounced, well outside of the range of living humans (compare A.L. 333-63 and human proximal hand phalanges shown in inset on the following page). Study of the tarsal-metatarsal joint articulation of the first or great toe of *A. afarensis* (A.L. 333-28,54), however, shows lack of grasping capabilities seen in all extant apes (Latimer and Lovejoy, 1990). It is unlikely, therefore, that these early hominids possessed the same level or kind of commitment to arboreality as the living apes. The curvature of the hand phalanges as well as the more cranial orientation of the glenoid fossa of the scapula suggest, however, that *A. afarensis* may have been a competent tree climber.

It is the consensus among paleoanthropologists that one differentiating (and perhaps most important) feature of hominids is their habitual and efficient bipedal locomotion. In addition to characteristics described above, there are other indicators of bipedality in the Lucy postcranial skeleton. In the drawing on page 52 (skeletons adjusted to same size), it can be seen that although the form of bipedalism may have been different from later hominids, the pelvis exhibits a number of morphological adaptations that are more similar to a modern human than to a quadrupedal ape, such as the chimpanzee. In particular, the ilium of Lucy is short and broad and flares outward, thus providing for muscle and bone positions compatible with habitual bipedality. In contrast to this configuration, the ape ilium is long and narrow. The angle formed at the knee region (bicondylar angle) in Lucy is also like that of the modern biped. In the chimpanzee, the angle is much straighter than in either *A. afarensis* or modern *Homo sapiens*. That is, the long axis of the femur is nearly parallel to the vertical axis; in modern humans, the long axis of the femur is not parallel to the vertical axis, but instead, forms a distinct angle. Note also the variation in position of the knees relative to each other in the chimpanzee, Lucy, and modern *Homo sapiens*. The knees of the chimpanzee are much further apart than those of either of the two bipeds. These anatomical details, coupled with footprint patterns from Laetoli (White and Suwa, 1987), provide irrefutable proof that these early hominids were bipedal.

References: Johanson and Edey, 1981; Johanson, Lovejoy, Kimbel, White, Ward, Bush, Latimer, and Coppens, 1982; Bush, Lovejoy, Johanson, and Coppens, 1982; Skinner and Sperber, 1982; Cherfas, 1983; Cook, Buikstra, DeRousseau, and Johanson, 1983; Jungers, 1982, 1988; Jungers and Stern, 1983; Stern and Susman, 1983; Susman, Stern, and Jungers, 1984, 1985; Wolpoff, 1983a, 1983b, 1998; Lovejoy, 1974, 1981, 1984, 1988; Lewin, 1983a, 1983b, 1987; Lamy, 1983; Berge, Orban-Segebarth, and Schmid, 1984; Tague and Lovejoy, 1986; McHenry, 1986; White and Suwa, 1987; Ruff, 1988; Tuttle, 1988; Latimer and Lovejoy, 1989, 1990; Simons, 1989; Ruff, 1991; Skelton and McHenry, 1992; Gebo, 1992, 1996; Häusler and Schmid, 1995; Simpson, 1996.

A.L. 288-1

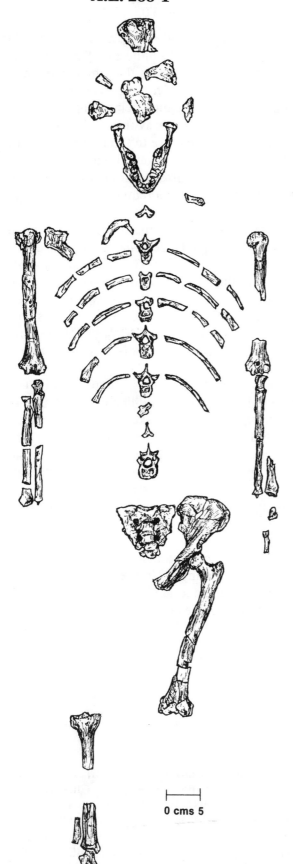

0 cms 5

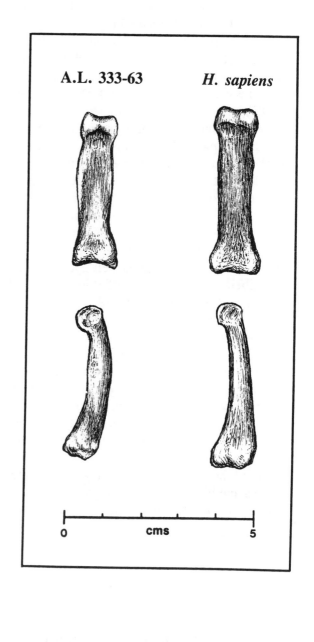

A.L. 333-63   *H. sapiens*

0   cms   5

50

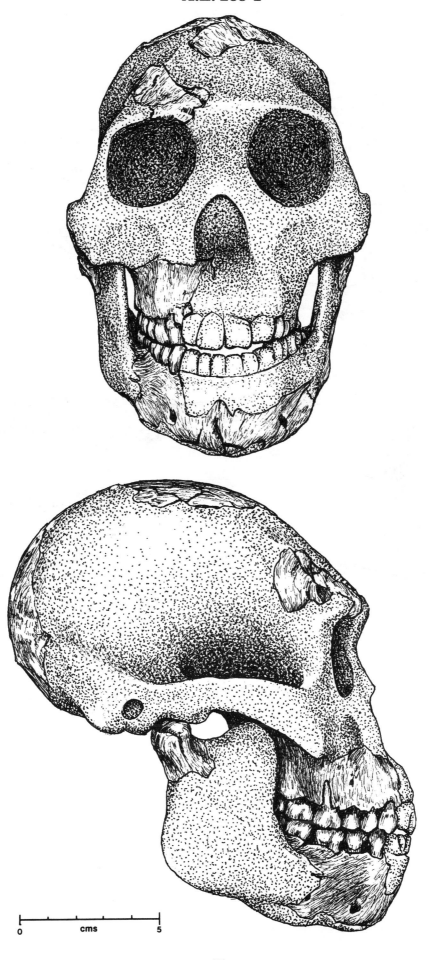

0 cms 5

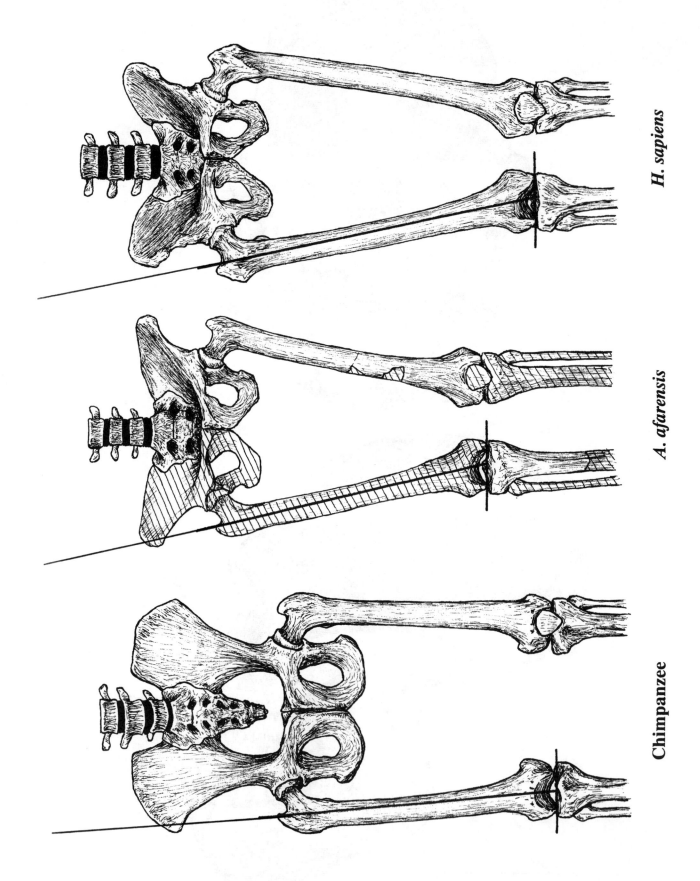

H. sapiens

A. afarensis

Chimpanzee

A.L. 200-la, A.L. 400-la

Specimen: **A.L. 200-1a, A.L. 400-1a**

Taxonomic affiliation: *Australopithecus afarensis*

Geographic location: Hadar, Ethiopia

Dating: Middle Pliocene (3.0 myBP)

General description: These fossil specimens were recovered by expeditions (1974–1977) to the Afar Depression by D. C. Johanson and co-workers. A.L. 200-la is a partial maxilla with a full adult dentition. The incisors are wide, particularly the central incisors. The canines are large, project above the tooth row, and show wear on the back portion of the chewing surface. There is a distinct diastema between the incisor and canine. The rear of the tooth row (molars) is markedly convergent. A.L. 400-la consists of the left and right portions of the mandible minus the ascending rami and part of the base. At least some of the parts that are missing from this specimen are due to carnivore chewing. The dentition is complete except for the right central incisor. Like the maxillary teeth, the mandibular teeth have a number of primitive, ape-like characteristics. Note, for example, the large, dominant cusp on the first premolar and the overall shape of the tooth row.

A.L. 200-la **A.L. 400-la**

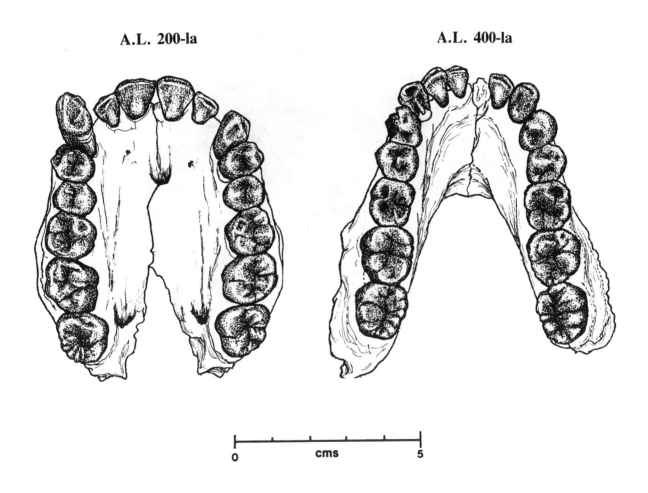

0 cms 5

References: Johanson and Edoy, 1981; Johanson and White, 1979; White and Johanson, 1982; Johanson, White, and Coppens, 1978, 1982; Kimbel, Johanson, and Coppens, 1982; Greenfield, 1990.

Hadar cranial reconstruction

Specimen: Hadar cranial reconstruction
Geographic location: Hadar, Ethiopia

Taxonomic affiliation: *Australopithecus afarensis*
Dating: Middle Pliocene (3.0 myBP)

General description: Originally reconstructed by T. White from 13 fossil specimens from a number of individuals, and subsequently modified by W. Kimbel and White (1988), this composite skull shows important anatomical features of *Australopithecus afarensis*. Note the great forward and downward placement of the lower face that is dominated by large front teeth and large, curved roots. Note also the presence of the upper diastema (between the lateral incisor and canine) and the lower diastema (between the canine and first premolar). According to this reconstruction, the upper face is small and chimpanzee-like. The cranium in general combines a small brain, large masticatory apparatus, and massive development of cranial base bone. This reconstruction is especially important in that it presents us with a graphic representation of the distinguishing features of the earliest known hominid. Compare the reconstruction with the relatively complete cranium (see A.L. 444-2).

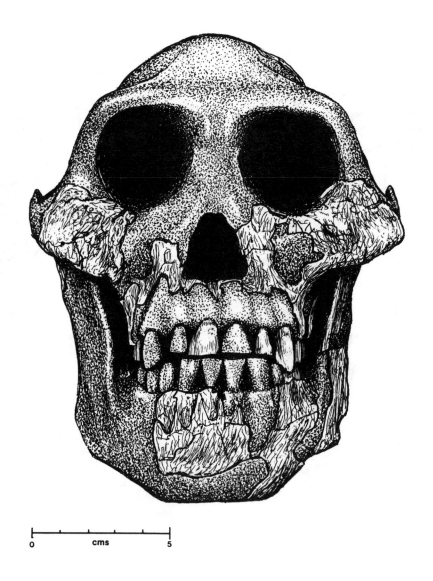

0   cms   5

References: Johanson, 1976; Johanson and Taieb, 1976; Johanson, White, and Coppens, 1978; Johanson and White, 1979; Johanson and Edey, 1981; White, Johanson, and Kimbel, 1981; Kimbel, Johanson, and Coppens, 1982; Kimbel, White, and Johanson, 1984; Lewin, 1982; Rak, 1983; Kimbel and White, 1988; Kimbel, Johanson, and Rak, 1994.

54

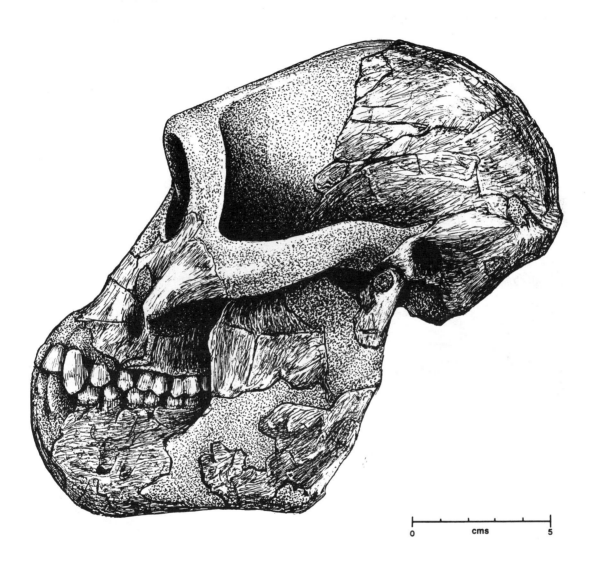

0   cms   5

A.L. 444-2

Specimen: **A.L. 444-2**
Geographic location: Hadar, Ethiopia

Taxonomic affiliation: *Australopithecus afarensis*
Dating: Middle Pliocene (3.0 myBP)

General description: This adult male cranium was found by Yoel Rak in 1992 as part of an expedition headed by Donald Johanson. It was subsequently reconstructed and described by William Kimbel and co-workers. This is the only relatively complete cranium of *Australopithecus afarensis*. The canines are massive, and the anterior teeth are much larger relative to the posterior teeth than in later hominids. The reconstruction is consistent with the Hadar composite reconstruction, confirming that the cranial variation seen at Hadar is likely intraspecific and not due to the presence of multiple hominid taxa at the site. The frontal bone is similar to *A. afarensis* recovered from Belohdelie dating to 3.9 myBP. This suggests the long existence (and perhaps stasis) of *A. afarensis*, approaching—if not exceeding—one million years.

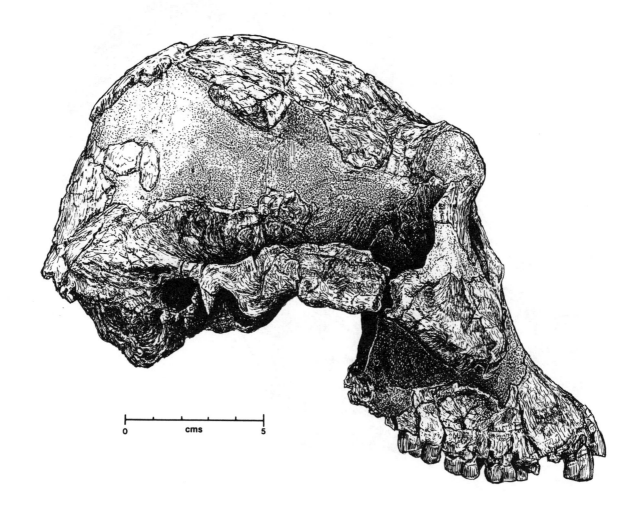

0   cms   5

References: Kimbel, Johanson, and Rak, 1994; Shreeve, 1994; White, Suwa, Hart, and others, 1993; Johanson and Blake, 1996; Wolpoff, 1998.

KNM-WT 17000

Specimen: **KNM-WT 17000**
Geographic location: Lomekwi,
  West Turkana, Kenya

Taxonomic affiliation: *Australopithecus aethiopicus*
Dating: Upper Pliocene (2.5 myBP)

General description: This virtually complete, undistorted cranium—known as the "Black Skull"—was discovered by Alan Walker while working with a field party from the National Museums of Kenya in 1985. The completeness of the specimen, coupled with recent refinements in chronology of the region, extends back in time the early record of these remarkably robust hominids. The hyper-robusticity of this specimen is quite similar to that of *A. boisei*. Note especially the very large face and the presence of a well developed sagittal crest (the largest known in the Hominidae). This fossil shows that an adaptation involving extremely heavy chewing was not unique to the latest australopithecines (e.g., KNM-ER 406, O.H. 5) as had been thought until quite recently by some workers. Observations of facial morphology and other characteristics by Kimbel and co-workers (1988) suggest that it warrants placement in a separate species, *A. aethiopicus*, a taxon originally defined by Arambourg and Coppens (1968) based on finds from the Omo River in Ethiopia. The brain is very small (410 cc). Only two other australopithecines—both from Hadar—have smaller brains.

```
|————|————|————|————|————|
0         cms            5
```

References: Arambourg and Coppens, 1968; Walker, R. Leakey, Harris, and Brown, 1986; Walker and R. Leakey, 1988; R. Leakey and Walker, 1988; Bower, 1987; Delson, 1986b; Lewin, 1986; Shipman, 1986; Johanson and White, 1986; Clark, 1988; Kimbel, White, and Johanson, 1988; Holloway, 1988; Grine, 1988; Falk, 1987, 1988; Tuttle, 1988; Simons, 1989; Suwa, White, and Howell, 1996; Wolpoff, 1998.

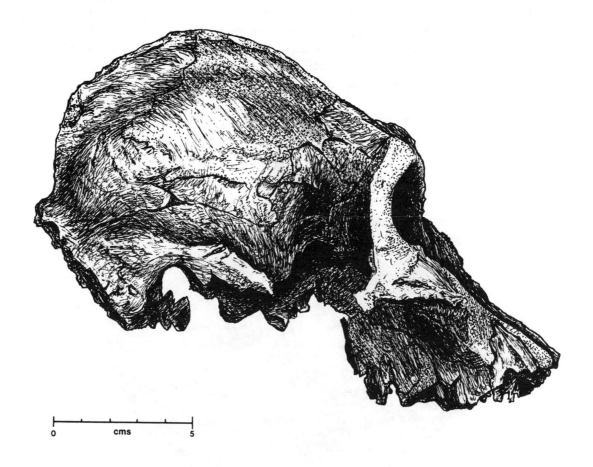

0   cms   5

Taung

Specimen: **Taung**
Geographic location: Taung, Republic of
  South Africa

Taxonomic affiliation: *Australopithecus africanus*
Dating: Upper Pliocene (2.5 myBP)

General description: Comprised of a complete face, the dentition, and most of a natural endocranial cast, this individual was the first australopithecine cranium to be found (1924). Given the amalgam of ape-like and human-like features, Raymond Dart (1925) concluded that the child's skull from Taung was from "an extinct race of apes intermediate between living anthropoids and man." Several hallmark features of the Hominidae that Dart recognized include small, flat-wearing canines, a foramen magnum at the bottom of the cranium (indicating bipedality), and a human-like, albeit tiny, brain. Work by Dean Falk (see Falk, 1980, 1983b, 1984) suggests that the brain may have been more ape-like than human-like (but see Holloway, 1984). Comparison of the dental development of this individual with standards based on modern human populations suggests that this individual was about six years of age at death (Mann, 1975, 1988). Application of three-dimensional computer imaging technology has made it possible to make precise observations on the unerupted permanent dentition as well as other hidden anatomical structures of this individual (Conroy, 1987; Conroy and Vannier, 1987). This work indicates that the Taung child was more ape-like in dental maturation and may have been only three to four years old at the time of death. In total, this individual shows a mosaic of ape-like and human-like characteristics.

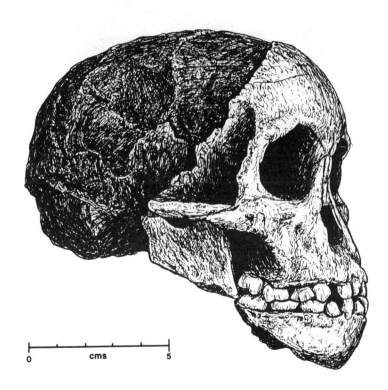

0	cms	5

References: Dart, 1925, 1929, 1967; Robinson, 1956; Tobias, 1971, 1974, 1984, 1985; Tobias and Falk, 1988; Howell, 1978; Falk, 1980, 1983b, 1984, 1989; Johanson and Edey, 1981; Reed, 1983; Rak, 1983; Holloway, 1984; Lewin, 1985, 1987; Howells, 1985: Vogel, 1985; Bromage, 1985; Grine, 1985; Conroy, 1987; Conroy and Vannier, 1987, 1989; Mann, 1975, 1988; Wolpoff, Monge, and Lampl, 1988; Tuttle, 1988; Falk, Hildebolt, and Vannier, 1989; McKee, 1993; McKee and Tobias, 1994; Wolpoff, 1998.

STS 5

Specimen: **STS 5**
Geographic location: Sterkfontein, Republic of South Africa

Taxonomic affiliation: *Australopithecus africanus*
Dating: Upper Pliocene (2.3–2.8 myBP)

General description: Discovered in 1947 by Robert Broom and John T. Robinson, this nearly complete adult cranium (nicknamed "Mrs. Ples"), is the most complete australopithecine cranium. This specimen shows a number of important cranial morphological features that are associated with *A. africanus*: small endocranial capacity (about 450 cc), lightly constructed cranium, reduced postorbital constriction, lower facial projection, wide zygomas, and relatively wide, dished face.

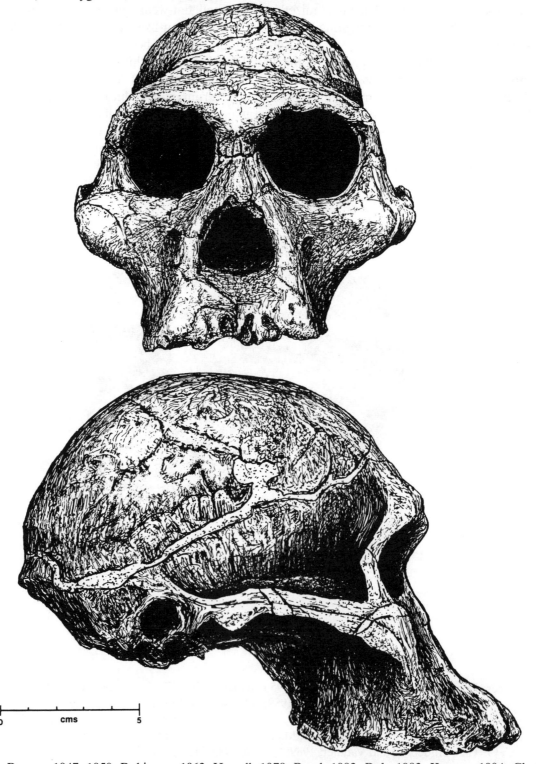

References: Broom, 1947, 1950; Robinson, 1962; Howell, 1978; Reed, 1983; Rak, 1983; Kuman, 1994; Clarke, 1994; Conroy, 1997.

Specimen: **STS 14**
Geographic location: Sterkfontein, Republic of South Africa

Taxonomic affiliation: *Australopithecus africanus*
Dating: Upper Pliocene (2.3–2.8 myBP)

General description: This partial postcranial skeleton (including vertebrae, ribs, sacrum, innominates, femur) was found by R. Broom and J. T. Robinson in 1947. It is an important set of materials because it shows skeletal evidence for bipedality in the australopithecines. The general configuration of the femur as well as the short and very broad ilium are clear indicators of this form of locomotion. The femoral head is small and the femoral neck is long compared to that of living humans. This skeleton is the first fossil material that demonstrated the antiquity of pre-*Homo* bipedality. Study of a recently recovered partial skeleton dating to about 3.2 myBP at Sterkfontein (STS 431) suggests that this hominid may have spent a significant amount of time in the trees.

References: Broom, Robinson, and Schepers, 1950; Lovejoy, 1976, 1988; Kuman, 1994; Clarke, 1994; Häusler and Schmid, 1995; Shreeve, 1996; Gebo, 1996; Conroy, 1997.

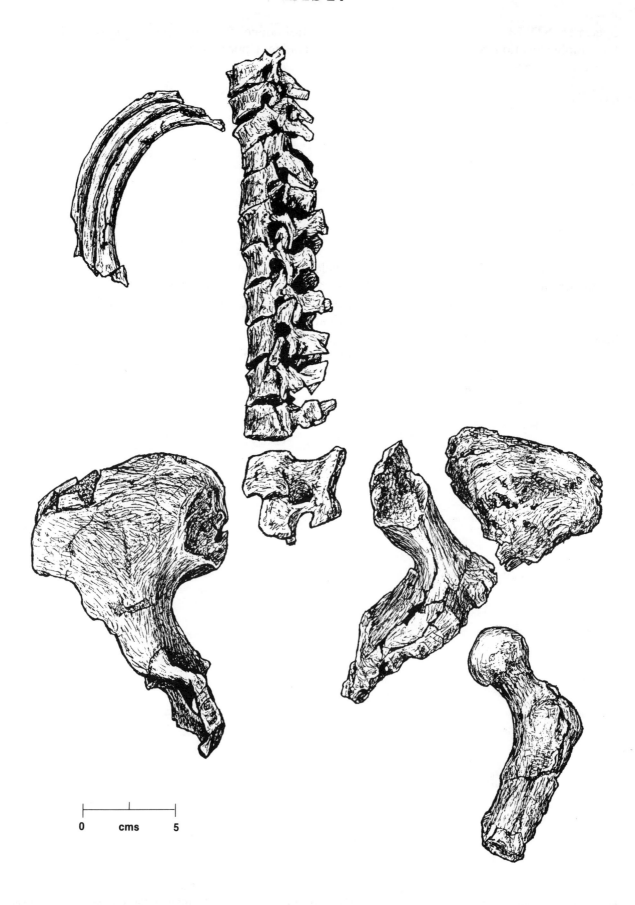

0   cms   5

STS 52a, STS 52b

Specimen: **STS 52a, STS 52b**
Geographic location: Sterkfontein, Republic
of South Africa

Taxonomic affiliation: *Australopithecus africanus*
Dating: Upper Pliocene (2.3–2.8 myBP)

General description: This is most of a maxilla (STS 52a) and mandible (STS 52b) with the full dentition recovered by J. T. Robinson in 1949. In general, the teeth and supporting bone show a disappearance of the obvious primitive features (e.g., larger canines and possession of diastema) that were seen in the earlier *Australopithecus afarensis*. The dentition is larger than earlier australopithecines, suggesting an increase in emphasis on the masticatory apparatus in this taxon of Hominidae.

References: Broom, Robinson, and Schepers, 1950; Robinson, 1954, 1956; Dart, 1954, 1962; White, Johanson, and Kimbel, 1981; Skinner and Sperber, 1982; Kuman, 1994; Clarke, 1994; Conroy, 1997.

STS 52a

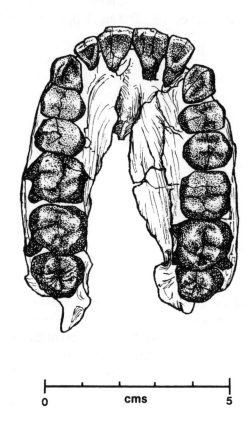

0 | cms   5

STS 52b

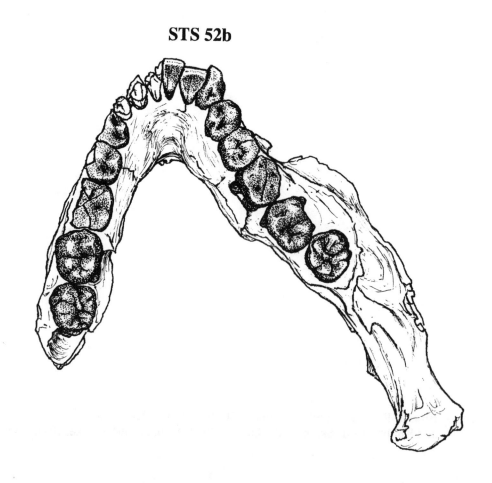

Specimen: **STS 71**
Geographic location: Sterkfontein, Republic of South Africa

Taxonomic affiliation: *Australopithecus africanus*
Dating: Upper Pliocene (2.3–2.8 myBP)

General description: Among the oldest of the South African australopithecines, this specimen was recovered by R. Broom and J. T. Robinson in 1947. It is missing most of the left side of the vault, but is, nevertheless, one of the more complete crania from Sterkfontein. The face shows forward projection, the temporal lines are positioned high on the cranium, indicating a large development of the chewing muscles. The brain is quite small (428 cc). Although primitive, this cranium is not as primitive or chimpanzee-like as that of *Austrulopithecus afarensis* (see Hadar cranial reconstruction).

References: Broom, Robinson, and Schepers, 1950; Holloway, 1975; White, Johanson, and Kimbel, 1981; Rak, 1983; Kuman, 1994; Clarke, 1994; Conroy, 1997; Wolpoff, 1998.

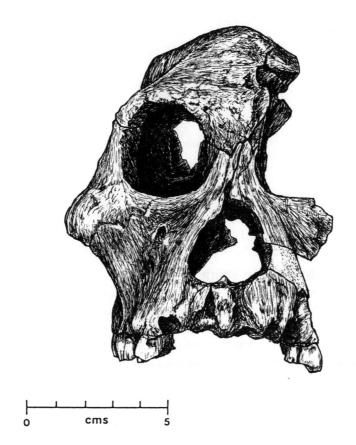

0 | cms | 5

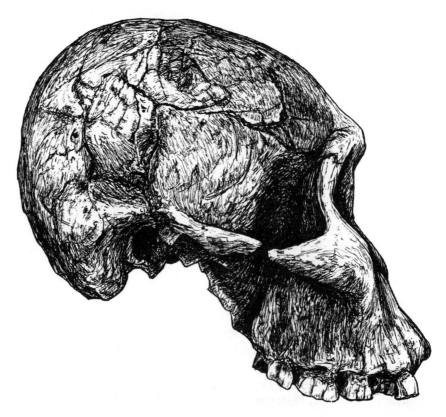

SK 46

Specimen: **SK 46**

Geographic location: Swartkrans, Republic of South Africa

Taxonomic affiliation: *Australopithecus robustus*

Dating: Late Pliocene—early Pleistocene (1.5–2.0 myBP)

General description: This individual was recovered along with a series of other important specimens from Swartkrans by Robert Broom and J. T. Robinson during the 1949–1952 field seasons. Although the specimen is badly crushed with only the left half of the cranial vault, portions of the face and palate, and a partial dentition present, a number of important features indicate an individual with a powerful masticatory apparatus—namely, well developed sagittal crest and robust face with an especially wide zygomatic region. Robinson (1960) noted that O.H. 5 and this australopithecine from Swartkrans showed similar patterns of craniofacial anatomy, thus establishing the presence of a form of robust australopithecine in both east and south Africa.

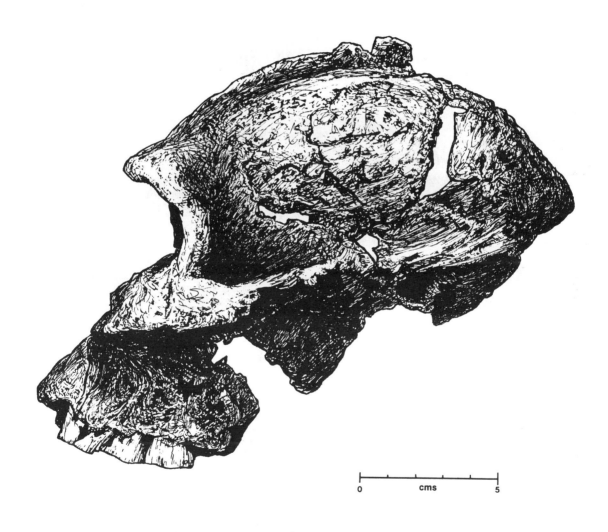

0   cms   5

References: Broom, 1949; Broom and Robinson, 1952; Robinson, 1956, 1960; Howell, 1978; Reed, 1983; Rak, 1983; Grine, 1988; Brain, Churcher, Clark, and others, 1988; Brain, 1993; Wolpoff, 1998.

SK 23, SK 48

Specimen: SK 23, SK 48
Geographic location: Swartkrans, Republic of
South Africa

Taxonomic affiliation: *Australopithecus robustus*
Dating: Late Pliocene—early Pleistocene (1.5–2.0
myBP)

General description: Both specimens were recovered during mining activities at the excavation site. In spite of the breakage of the cranium (SK 48) due to blasting, R. Broom and J. T. Robinson were able to piece back together most of the fragments (Broom and Robinson, 1952). The cranium is distorted, but shows important characteristics that are typical of the taxon of *A. robustus*—sagittal crest, well developed face, and large posterior teeth. Note the very wide zygomatic region, and in the mandible, the wide ascending ramus. The extreme of these features is exhibited in the earlier *A. aethiopicus* (see KNM-WT 17000) and the later *A. boisei* (see KNM-ER 406, KNM-ER 732, O.H. 5). The mandible (SK 23) was minimally affected by blasting, but it is deformed. In particular, the two halves have been squeezed together, thus resulting in an abnormally narrow configuration. In spite of this deformation, the mandible is extremely well preserved with the entire permanent dentition present. Overall, the mandible is very robust and, although from a different individual, it is probably not too dissimilar to the one that belonged to SK 48 in actual life.

SK 48

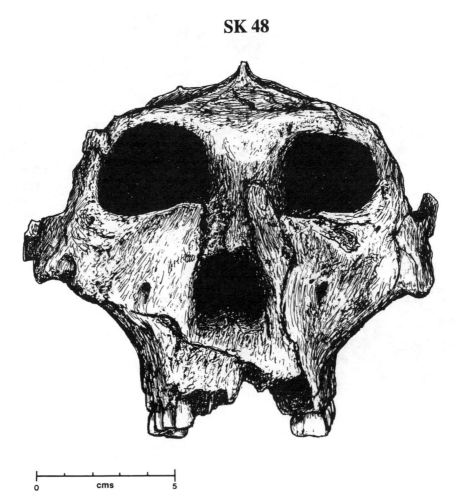

0   cms   5

References: Broom and Robinson, 1950; Robinson, 1956, 1960; Skinner and Sperber, 1982; Reed, 1983; Rak, 1983; Grine, 1988; Brain, Churcher, Clark, and others, 1988; Brain, 1993.

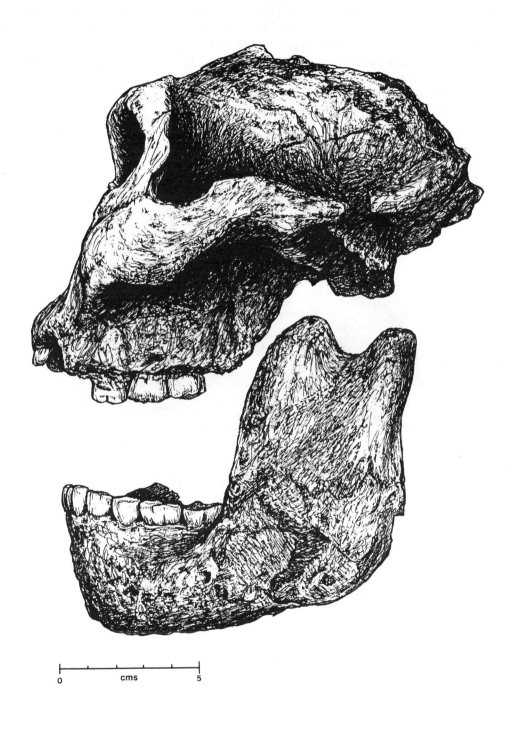

0   cms   5

O.H. 5

Specimen: **O.H. 5**
Geographic location: Olduvai Gorge, Tanzania

Taxonomic affiliation: *Australopithecus boisei*
Dating: Late Pliocene—early Pleistocene
(1.8 myBP)

General description: This nearly complete cranium (colloquially known as "Zinj" after the original genus name, *"Zinjanthropus"*) was discovered by Mary Leakey in 1959 and subsequently described by P. Tobias. Both the size of the attachment areas for the masticatory muscles—note the well developed sagittal crest—and the molars are massive. In addition, the premolars are enlarged (molarized), effectively increasing the chewing surface of the cheek teeth from three to five teeth on each side of the jaw, upper and lower. Note the relatively tiny incisors and the flat occlusal wear pattern of the teeth, including the canines. The enamel is thick. The face is massive with wide zygomas, giving it a diamond shape. These features characterize this taxon of hominid. The method of dating of this specimen (potassium argon) was important in that it provided an extremely old date, and it represented the first reliable chronometric assessment of an early hominid.

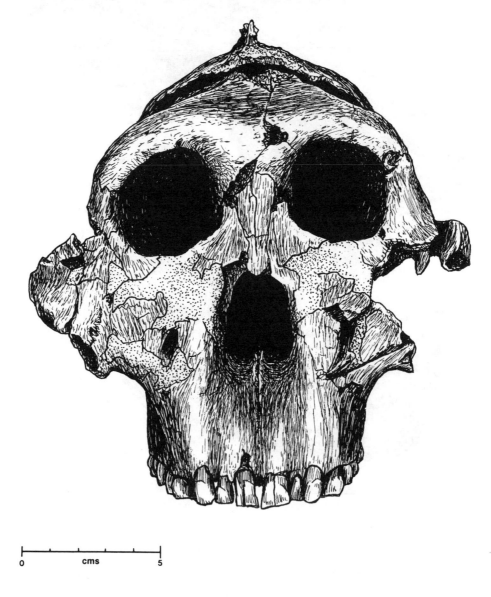

0   cms   5

References: L. Leakey, 1960; Robinson, 1960; Tobias, 1967; Johanson and Edey, 1981; Rak, 1983; M. D. Leakey, 1984; Grine, 1988; Demes and Creel, 1988.

O.H. 5

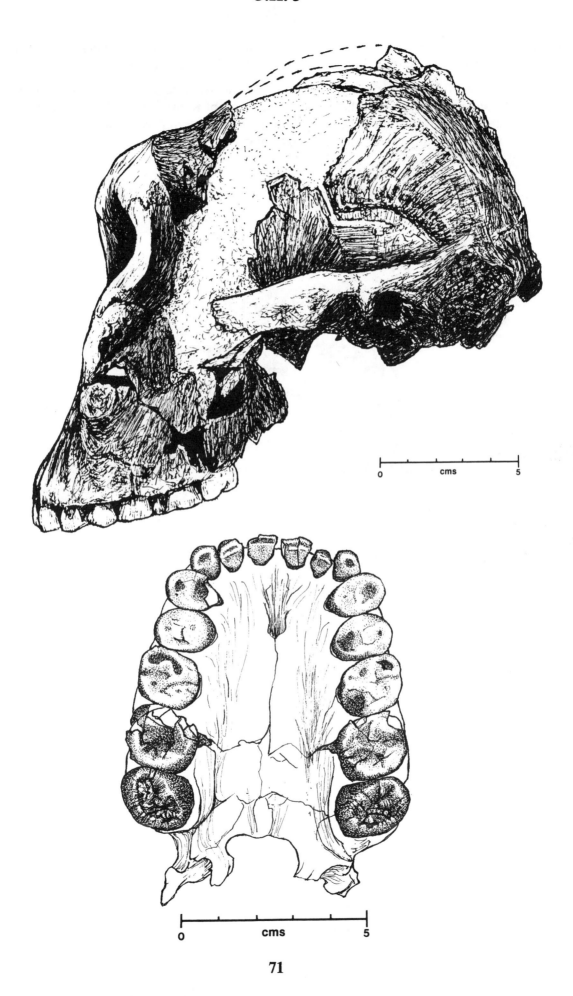

cms

0   5

cms

0   5

Peninj (Natron)

Specimen: **Peninj (Natron)**
Geographic location: Lake Natron, Tanzania

Taxonomic affiliation: *Australopithecus boisei*
Dating: Early Pleistocene (1.5 myBP)

General description: This is a nearly complete mandible and lower dentition that was discovered by K. Kimeu in 1964 under the auspices of the National Museums of Kenya. This individual shows the characteristic features of the hyper-robust australopithecines from eastern Africa. The dentition—especially the post-canine teeth—is huge and is supported by a massive mandible. Note also the very tiny incisors.

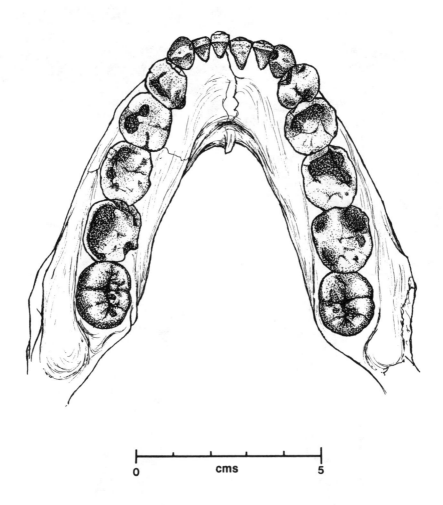

References: L. Leakey and M. D. Leakey, 1964; Tobias, 1965b; Grine, 1988; Wolpoff, 1998.

KNM-ER 406

Specimen: **KNM-ER 406**

Geographic location: Ileret, East Turkana, Kenya

Taxonomic affiliation: *Australopithecus boisei*

Dating: Early Pleistocene (1.5 myBP)

General description: This specimen is represented by a nearly complete cranium with no dentition. It was discovered by Richard Leakey in 1969. Like that of the O.H. 5 individual, this cranium exhibits well developed sagittal and nuchal (occipital) crests with large mastoid processes and a markedly robust, wide face. Unlike many of the earlier australopithecines (e.g., *A. africanus*), this taxon shows pronounced postorbital constriction. The endocranial capacity estimate for this individual—510 cc— reflects very little, if any, brain expansion in comparison of early and late australopithecines. Note the overall similarity of this cranium with KNM-WT 17000 and O.H. 5 specimens.

References: R. Leakey, 1970, 1971, 1973b, 1974, 1976; R. Leakey, Mungai, and Walker, 1971; Holloway, 1973; R. Leakey and Walker, 1976; R. Leakey, M. G. Leakey, and Behrensmeyer, 1978; Walker and R. Leakey, 1978; Rak, 1983; Grine, 1988; Wood, 1991.

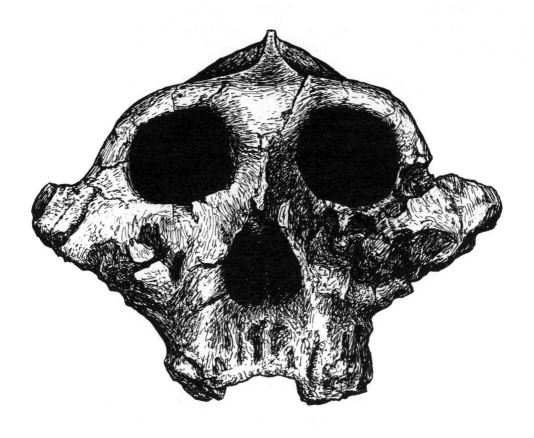

0   cms   5

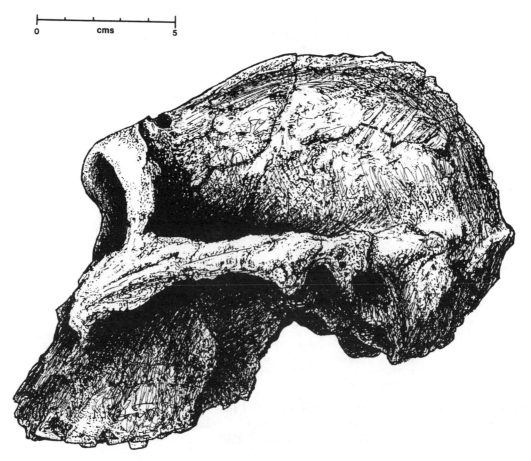

KNM-ER 732

Specimen: **KNM-ER 732**   Taxonomic affiliation: *Australopithecus boisei*
Geographic location: Ileret, East Turkana, Kenya   Dating: Early Pleistocene (1.5 myBP)

General description: This partial cranium was discovered by M. Mutua in 1970 with the Koobi Fora
Research Project under the directorship of R. Leakey. The specimen consists of most of the right side
of the face and top and right side of the vault. The endocranial capacity of the individual is small (500cc).
The postorbital constriction is quite marked and the overall configuration suggests that this may be a
female counterpart to a male *A. boisei* (e.g., KNM-ER 406, O.H. 5).

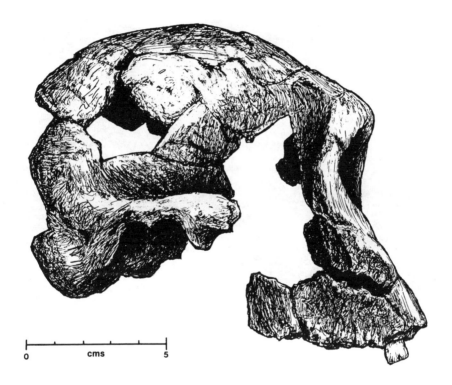

0   cms   5

References: R. Leakey, 1971, 1974; R. Leakey, Mungai, and Walker, 1971; R. Leakey, M. G. Leakey, and Behrensmeyer, 1978; Holloway, 1973; Walker and R. Leakey, 1978; Johanson and White, 1979; Rak,
1983; Grine, 1988; Wood, 1991; Wolpoff, 1998.

KNM-ER 1805

Specimen: **KNM-ER 1805**

Geographic location: Koobi Fora, East Turkana, Kenya

Taxonomic affiliation: *Australopithecus sp.*

Dating: Early Pleistocene (1.5–1.6 myBP)

General description: This specimen was discovered by P. Abell of the Koobi Fora Research Project in 1973. Shown here is the cranium; an associated mandible was also recovered. With the exception of the missing face and browridge area, the cranium is relatively complete. Some workers have emphasized characteristics associated with *Homo* in this individual (e.g., Wood, 1991; Wolpoff, 1998). However, the presence of a sagittal crest, generally robust cranium, small brain (less than 600 cc.), and ape-like sulcal pattern as seen in the endocast suggests that this specimen is more closely aligned with the taxon *Australopithecus* (see Falk, 1983a).

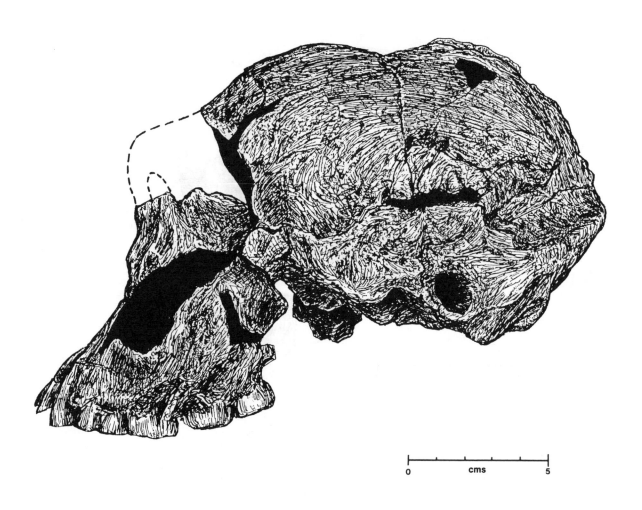

0   cms   5

References: R. Leakey, 1974; Day, R. Leakey, Walker, and Wood, 1976; Howell, 1978; Walker, 1981; Falk, 1983a; Wood, 1991; Wolpoff, 1998.

Homo habilis

KNM-ER 1470

4. *HOMO HABILIS*

Pliocene-Pleistocene Hominids

Beginning between two and three million years ago, a new hominid appears in eastern and southern Africa. At the suggestion of Raymond Dart, Louis Leakey and colleagues Philip Tobias and John Napier chose the species name *"habilis"* after the Latin term for "able, handy, mentally skillful, vigorous." In comparison with australopithecines, this taxon has an appreciably larger brain, smaller cheek teeth, and in general lacks the specializations that are associated with a heavily masticated diet. Arm and leg bones of *Homo habilis* from Olduvai Gorge indicate that this taxon likely retained primitive characteristics of small body size and long arms (Johanson and others, 1987; Johanson and Shreeve, 1989). Precise dating indicates that this taxon is clearly a contemporary of the australopithecines. Study of hand bones from Olduvai Gorge suggests that the use of the hand in manipulating objects—possibly involving tools—may have been more precise than in earlier hominids. Until recently, the earliest dates on *Homo habilis* were well after 2 myBP. New materials attributed to this taxon from Hadar (maxilla and associated dentition) extend the dates to 2.3 myBP, well into the Pliocene (Kimbel and others, 1996, 1997). This taxon disappeared from Africa somewhat less than 1.6 million years ago, serving as the probable ancestor for *Homo erectus*.

The representatives of this taxon that we include in the following descriptions are:
 1. Olduvai Gorge, Tanzania: O.H. 13, O.H. 24, O.H. 8
 2. East Turkana, Kenya: KNM-ER 1813, KNM-ER 1470
 3. Sterkfontein, South Africa: StW 53
 4. Swartkrans, South Africa: SK 847

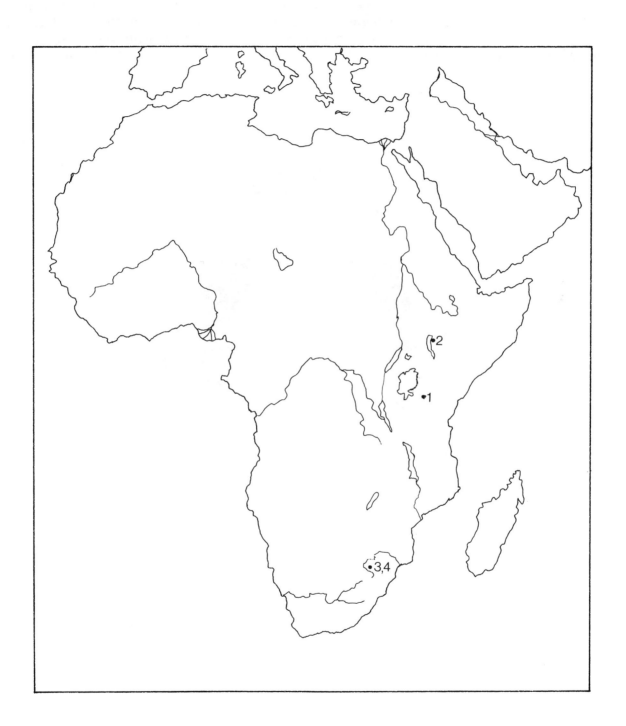

O.H. 13

Specimen: **O.H. 13**
Geographic location: Olduvai Gorge, Tanzania

Taxonomic affiliation: *Homo habilis*
Dating: Early Pleistocene (1.7 myBP)

General description: This hominid was discovered by N. Mbuika during a 1963 expedition headed by L. Leakey. Shown here is the occlusal view of the mandibular dentition. In addition to the mandible, parts of the maxilla, occipital, parietals, temporals, and frontal were recovered. Estimates of endocranial capacity and measurement of the masticatory complex—the teeth, in particular—show respective increase in brain size and decrease in size of the chewing apparatus in comparison with the australopithecines.

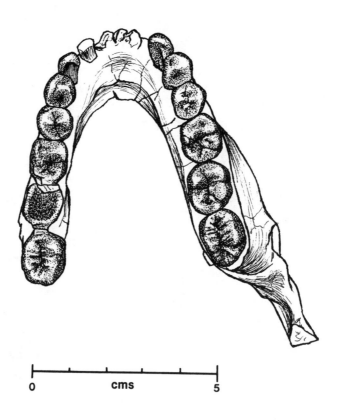

References: L. Leakey and M. D. Leakey, 1964; L. Leakey, Tobias, and Napier, 1964; Tobias and von Koenigswald, 1964; Robinson, 1965; Tobias, 1965b; Holloway, 1973; Brace, Mahler, and Rosen, 1973; Johanson and White, 1979; Skinner and Sperber, 1982; Tobias, 1991.

O.H. 24

Specimen: **O.H. 24**
Geographic location: Olduvai Gorge, Tanzania

Taxonomic affiliation: *Homo habilis*
Dating: Late Pliocene—early Pleistocene (1.75–2.0 myBP)

General description: Discovered by P. Nzube in 1968, the specimen had been exposed and badly weathered for a number of years. Restoration and reconstruction by R. J. Clarke yielded a specimen consisting of the better part of a cranial vault, and part of the face and dentition is present as well. Although some paleoanthropologists include it in the australopithecines, most agree that the presence of reduced postorbital constriction, a lighter-built cranium, expanded brain size (about 600 cc), and generally smaller teeth align it most closely with the genus *Homo*.

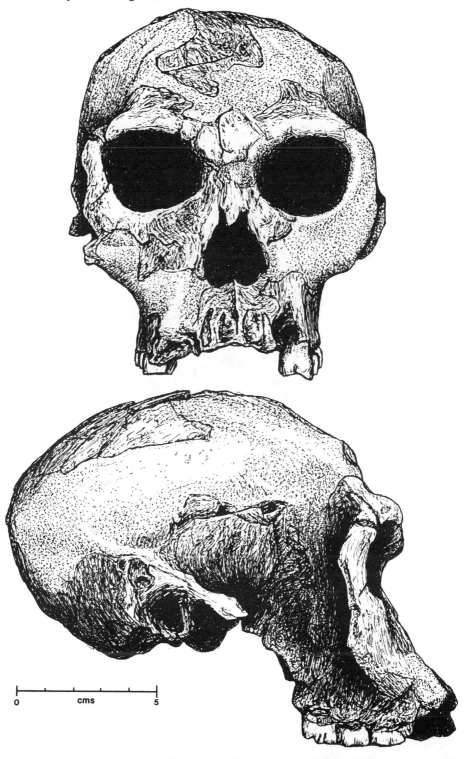

0   cms   5

References: M. D. Leakey, Clarke, and L. Leakey, 1971; Tobias, 1972, 1991; Howell, 1978.

O.H. 8

Specimen: **O.H. 8**
Geographic location: Olduvai Gorge, Tanzania

Taxonomic affiliation: *Homo habilis*
Dating: Late Pliocene—early Pleistocene (1.75 myBP)

General description: The O.H. 8 foot, discovered by an assistant to L.S.B. Leakey in 1960, belongs to a 13- or 14-year-old subadult. Most of the bones of a left foot are represented, lacking only the phalanges and the heel portion of the calcaneus. Features of this foot are very similar to those of modern humans, including the close alignment of the first metatarsal (big toe) beside the second metatarsal and the restricted mobility of the foot joints. This foot is clearly associated with human-like bipedality.

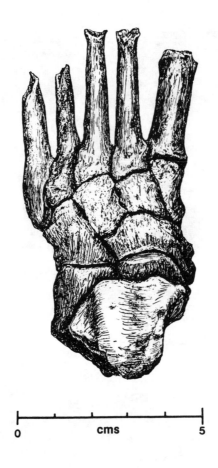

0   cms   5

References: L. Leakey, 1960; Day and Napier, 1964; Napier, 1967; Day and Wood, 1968; Lovejoy, 1975, 1976, 1978; Oxnard and Lisowski, 1980; Lamy, 1983; McHenry, 1984; Susman and Stern, 1982; Susman, 1983; White and Suwa, 1987; Kidd, O'Higgins, and Oxnard, 1996.

KNM-ER 1813

Specimen: **KNM-ER 1813**
Geographic location: Koobi Fora, East Turkana, Kenya

Taxonomic affiliation: *Homo habilis*
Dating: Late Pliocene—early Pleistocene (1.8 myBP)

General description: This virtually complete cranium was discovered in 1973 by K. Kimeu under the auspices of the Koobi Fora Research Project. It consists of most of the cranium and part of the cranial base and dentition. Like other habilines, the cranium is lightly built, globular in shape, and shows only moderate postorbital constriction.

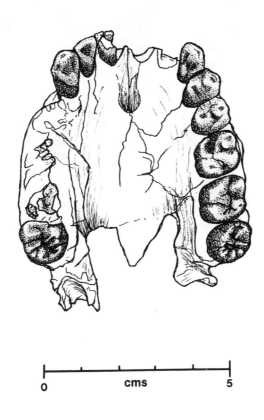

0   cms   5

References: R. Leakey, 1974, 1976; Day, R. Leakey, Walker, and Wood, 1976; Rak, 1983; Wood, 1991; Wolpoff, 1998.

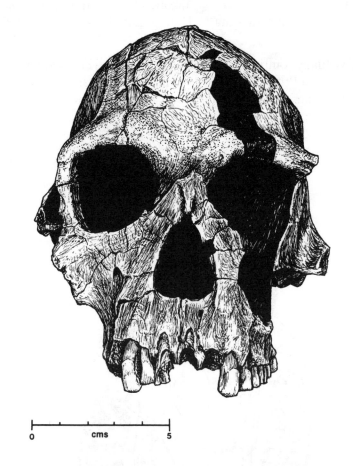

0   cms   5

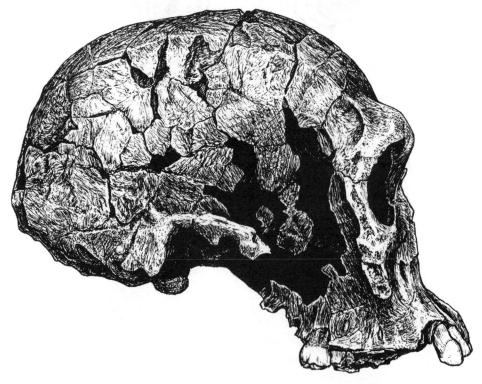

KNM-ER 1470

Specimen: **KNM-ER 1470**

Geographic location: Koobi Fora, East Turkana, Kenya

Taxonomic affiliation: *Homo habilis*

Dating: Late Pliocene—early Pleistocene (1.8 myBP)

General description: Discovered by B. Ngeneo under the team led by Richard Leakey (Koobi Fora Research Project) in 1972, the piecing together of the myriad of fragments of this cranium resulted in a reconstruction showing all of the features of *Homo habilis* that were only weakly revealed in other habiline specimens that had been recovered from Olduvai Gorge fossil localities (specimens: O.H. 7, O.H. 13, O.H. 16, O.H. 24). The important features of this rather large cranium include an endocranial volume of 775 cc, weakly developed muscle attachment sites, moderate postorbital constriction, and generally speaking, a lightly constructed, globular vault.

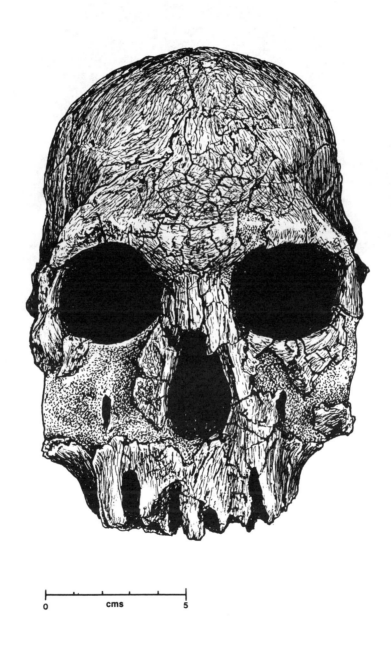

0   cms   5

References: R. Leakey, 1973a, 1973b; 1973c; R. Leakey and Walker, 1976; Wood, 1976; Rak, 1983; Falk, 1987; Wood, 1991.

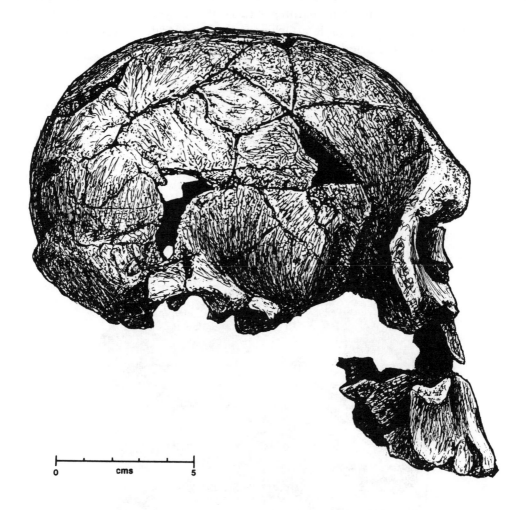

0 cms 5

StW 53

Specimen: **StW 53**

Geographic location: Sterkfontein, Republic of South Africa

Taxonomic affiliation: *Homo habilis*

Dating: Late Pliocene—early Pleistocene (1.5–2.0 myBP)

General description: Discovered in 1976 by A. R. Hughes of the University of the Witwatersrand (Republic of South Africa), this cranium consists of the face and fragments of the top and back as well as a partial dentition. The cranium is generally rounded and full with thin browridges, features, which in addition to others, distinguish habilines from australopithecines. This discovery established the presence of this early form of *Homo* in southern Africa.

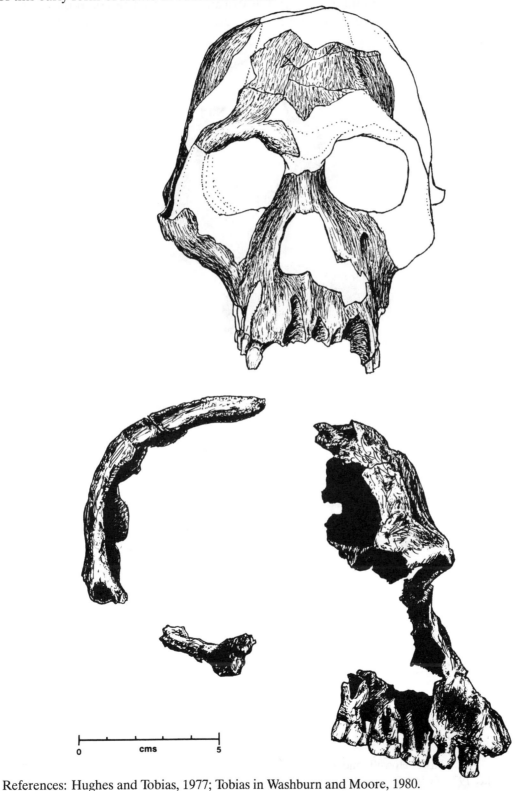

References: Hughes and Tobias, 1977; Tobias in Washburn and Moore, 1980.

SK 847

Specimen: **SK 847**

Geographic location: Swartkrans, Republic of South Africa

Taxonomic affiliation: *Homo (habilis?)*

Dating: Late Pliocene—early Pleistocene (1.5–2.0 myBP)

General description: Although this hominid was originally cataloged under several different numbers from the 1949–1952 field seasons, it was not recognized as a single individual until the late 1960s. The individual has been variously called *Homo* (Clarke and Howell, 1972; Clarke, 1977) and *Australopithecus* (Wolpoff, 1971, 1980) and has been the focus of discussion as to whether or not there was more than one species of contemporary hominids represented in the South African sequence. This composite cranium shows anatomical details that are closest to the morphological pattern associated with early *Homo*.

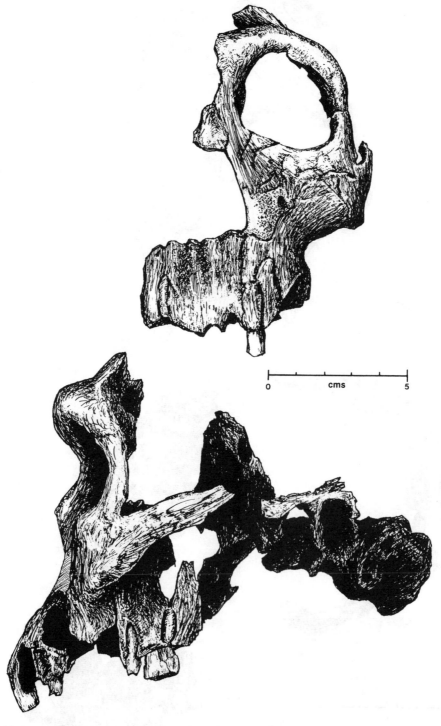

References: Robinson, 1953; Clarke, Howell, and Brain, 1970; Wolpoff, 1970, 1971, 1980; Clarke and Howell, 1972; Clarke, 1977; Grine, Demes, Jungers, and Cole, 1993; Brain, 1993.

Homo erectus

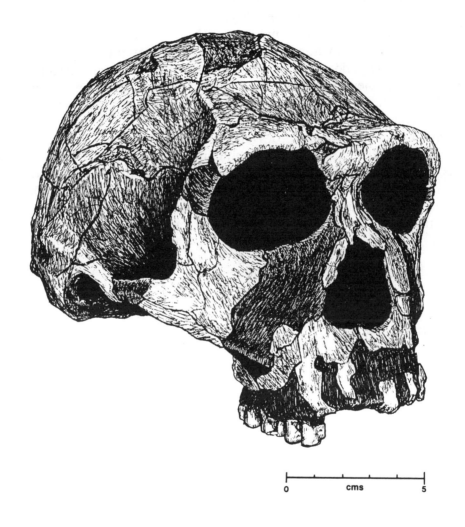

0   cms   5

KNM-ER 3733

5. HOMO ERECTUS

Pleistocene Hominids

Appearing about 1.8 million years ago in Africa is a new hominid, *Homo erectus*. This hominid probably descended from *Homo habilis* and was undoubtedly a contemporary of the last survivors of the australopithecines. The earliest representatives of *Homo erectus* have been found in east Africa to the immediate east and west of Lake Turkana (Koobi Fora and Nariokotome, respectively), and by one million years ago are present in Asia and possibly in Europe (Ackerman, 1989). The most recent forms of *Homo erectus* are dated at about 300,000 to 400,000 years ago. It is with this taxon, then, that we see the first movement of hominids out of Africa into previously unoccupied areas of the Old World. This expansion in area of occupation by hominids may have been related to major changes in adaptive strategy. It is during this time period, for example, that hunting of small game begins to become incorporated in resource procurement. It is doubtful, however, that large-scale hunting fully replaces scavenging activities. The appearance of a more elaborate technology (called Acheulian) evinces a more culturally complex hominid.

Relative to earlier hominids, the brain of *Homo erectus* is larger, the faces and cheek teeth are reduced in size, and the cranial bones are considerably thicker. Later *Homo erectus* has larger front teeth and larger brains than earlier *Homo erectus*. These changes indicate that the period of time associated with *Homo erectus* is not a period of stasis, but rather, gradual evolutionary change. On the other hand, other morphological features, such as relative thickness of the cranial vault bones as well as the limb bones, show very little change within the *Homo erectus* sequence.

The hominids and their associated site localities that we have chosen to represent for this period of evolution include the following:

1. East Turkana, Kenya: KNM-ER 3733, KNM-ER 3883
2. West Turkana, Kenya: KNM-WT 15000
3. Olduvai Gorge, Tanzania: O.H. 9
4. Bodo, Ethiopia
5. Ternifine, Algeria: Ternifine 2 and 3
6. Mauer, Germany
7. Sangiran, Indonesia: Sangiran 4, 17, IX
8. Hexian, China: PA 830
9. Zhoukoudian, China

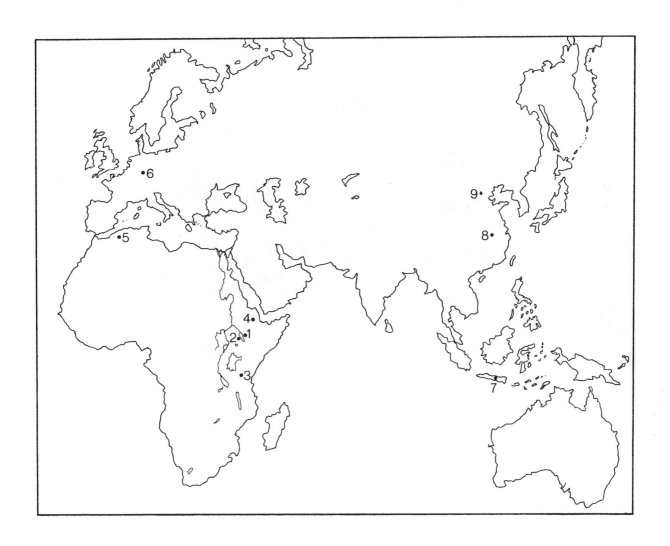

KNM-ER 3733

Specimen: **KNM-ER 3733**
Geographic location: Koobi Fora,
  East Turkana, Kenya

Taxonomic affiliation: *Homo erectus*
Dating: Early Pleistocene (1.8 myBP)

General description: Prior to the discovery of this cranium, the contemporaneity of at least two Plio-Pleistocene hominid taxa had been suspected by some and insisted upon by others. However, the finding of this nearly complete cranium in 1975 by a member of the Koobi Fora Research Project paleontological team in deposits of similar age as those associated with KNM-ER 406—a robust australopithecine—laid to rest a single lineage model of human evolution. The cranium shows an overall similarity to *Homo erectus* from Asia in that the endocranial volume has expanded (to 848 cc), and the postorbital constriction is greatly reduced. Likewise, the size of the posterior dentition, face, and attachment sites for the masticatory muscles are smaller than in the australopithecines. Note, too, the diminutive size of the browridges compared to other *H. erectus* and the sloping forehead.

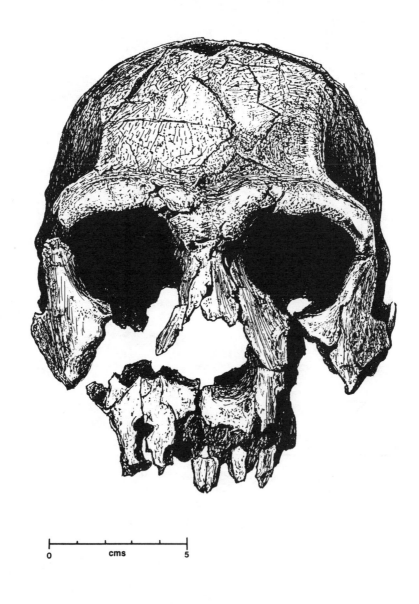

0   cms   5

References: R. Leakey, 1976; R. Leakey and Walker, 1976, 1985a; R. Leakey, M. G. Leakey, and Behrensmeyer, 1978; Walker and R. Leakey, 1978; Howell, 1978; Johanson and White, 1979; Johanson and Edey, 1981; Rightmire, 1990; Wolpoff, 1998.

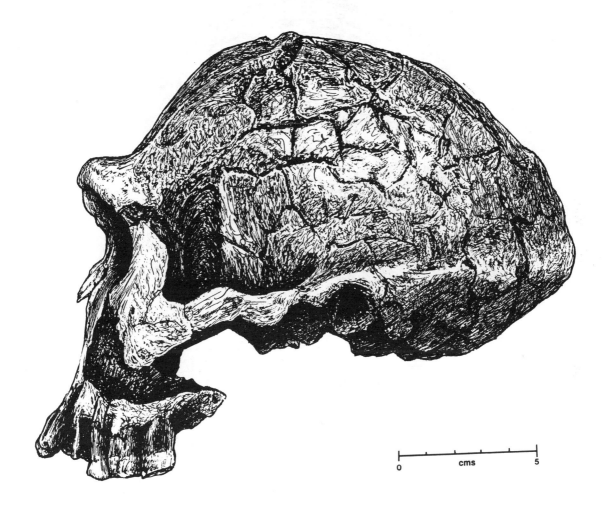

0   cms   5

Specimen: **KNM-ER 3883**
Geographic location: Koobi Fora,
  East Turkana, Kenya

Taxonomic affiliation: *Homo erectus*
Dating: Early Pleistocene (1.4–1.6 myBP)

General description: The specimen includes most of the cranium, but lacks much of the facial skeleton. Recovered by the Koobi Fora Research Project under the direction of R. Leakey, this specimen shows a great resemblance to KNM-ER 3733. Several features, however, show relatively greater size and development of this individual—browridges and preserved face, muscle attachment sites, mastoid processes, cranial base—leading one paleoanthropologist to suggest that this individual represents the male counterpart to the female KNM-ER 3733 (see Wolpoff, 1998). The endocranial capacity is on the same order as that recorded for KNM-ER 3733 (804 cc).

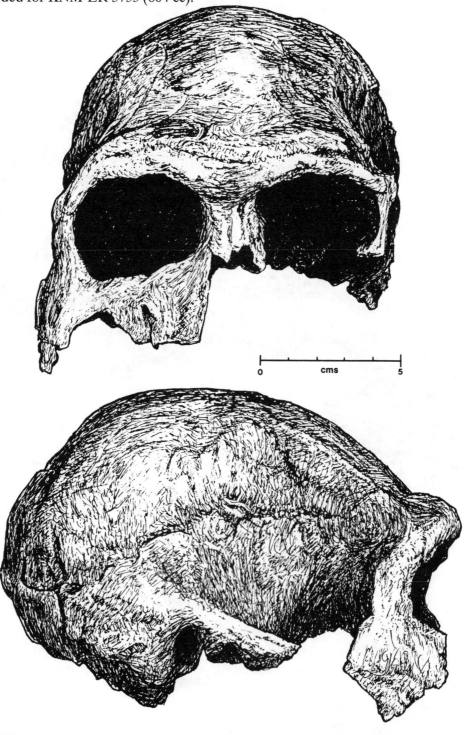

References: Walker and R. Leakey, 1978; Walker, 1981; R. Leakey and Walker, 1985a; Rightmire, 1990; Wolpoff, 1998.

KNM-WT 15000

Specimen: **KNM-WT 15000**
Geographic location: Nariokotome III,
  West Turkana, Kenya

Taxonomic affiliation: *Homo erectus*
Dating: Early Pleistocene (1.51–1.56 myBP)

General description: A cranial fragment of this individual was discovered by K. Kimeu of the National Museums of Kenya in 1984 during a survey of fossil deposits on the west side of Lake Turkana. The remainder of the bones were recovered by Alan Walker and Richard Leakey over the next three years. This hominid skeleton is the most complete (80% of the bones are present) until Neandertal times. Moreover, the date assigned to the specimen makes it the earliest *Homo erectus* found. The bones are exceedingly well preserved, having been subject to relatively little postmortem damage due either to geological processes or animal activity. Because of the completeness of the skull, dentition, and postcranium, a great deal of information is available from the study of this individual. For example, hominid growth patterns as well as the relationship between brain and body size have been at the foundation of much discussion regarding behavior in early hominids. The discovery of this individual makes it possible to begin to accurately measure these variables.

The robusticity and size of the cranial and postcranial bones as well as the morphology of the pelvis indicate that this individual was likely male. The stage of dental development suggests that this individual died between 11 and 15 years of age. The major epiphyses (ossification centers at the ends of the bones) are largely unfused, suggesting that more growth would have occurred in the individual—he would have been a little over six feet tall had he become fully mature. The brain size of the individual was slightly over 900 cc. Partially because of the immaturity of the individual, specific areas of the cranium, such as the attachment sites for the temporalis muscle, are not especially well developed. Had the individual reached adulthood, the vault may well have been as robust as some of the other *H. erectus* crania from Africa (e.g., O.H. 9). The pelvic region is similar to australopithecines in some respects, especially with regard to the wide flare of the ilium and the very long femoral neck. On the other hand, the birth canal appears small in relation to the brain volume, a modern characteristic. The relatively thick body trunk and long limbs of this individuals is not unlike those of modern African populations adapted to tropical settings. Thus, as to be expected, this individual expresses a mosaic of modern and archaic characteristics.

References: Brown, Harris, R. Leakey, and Walker, 1985; R. Leakey and Walker, 1985b, 1989; Huff, 1988; Simons, 1989; Smith, 1986, 1990; Rightmire, 1990; Ruff, 1991, 1993, 1994; Walker, 1993; Walker and Leakey, 1993.

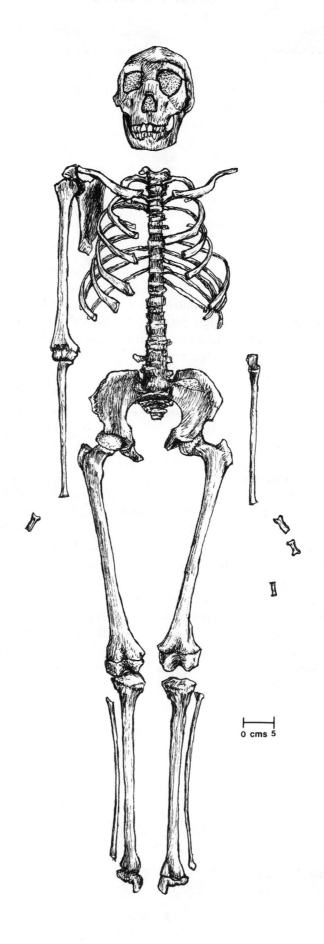

0 cms 5

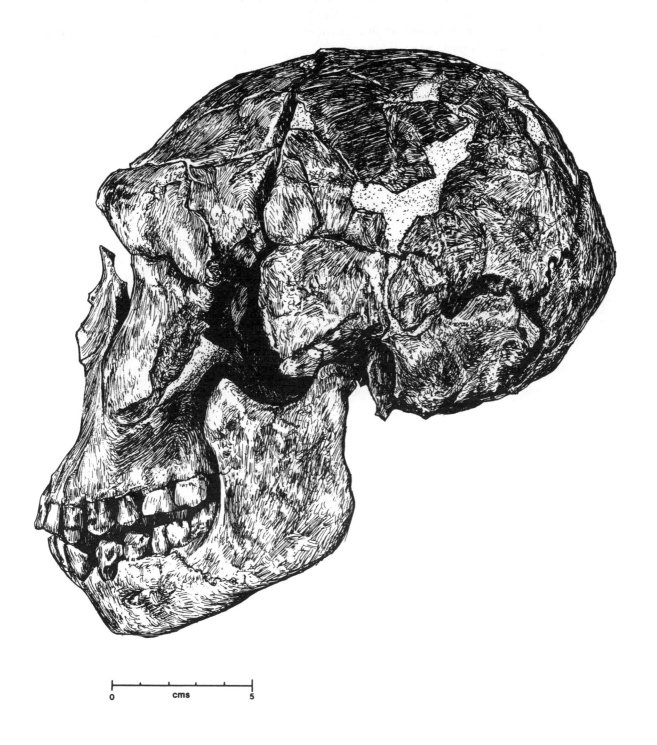

0   cms   5

O.H. 9

Specimen: O.H. 9

Geographic location: Olduvai Gorge, Tanzania

Taxonomic affiliation: *Homo erectus*

Dating: Early Pleistocene (1.2 myBP)

General description: Found by Louis Leakey in 1960 from Upper Bed II at Olduvai, this specimen consists of part of a cranial vault with the browridges and most of the base of the cranium. Features of this cranium are massive, including huge browridges, the largest known for any hominid. Like other *H. erectus* hominids, the cranium is heavily built and shows a marked expansion in endocranial volume (1067 cc) in comparison with those of earlier hominids.

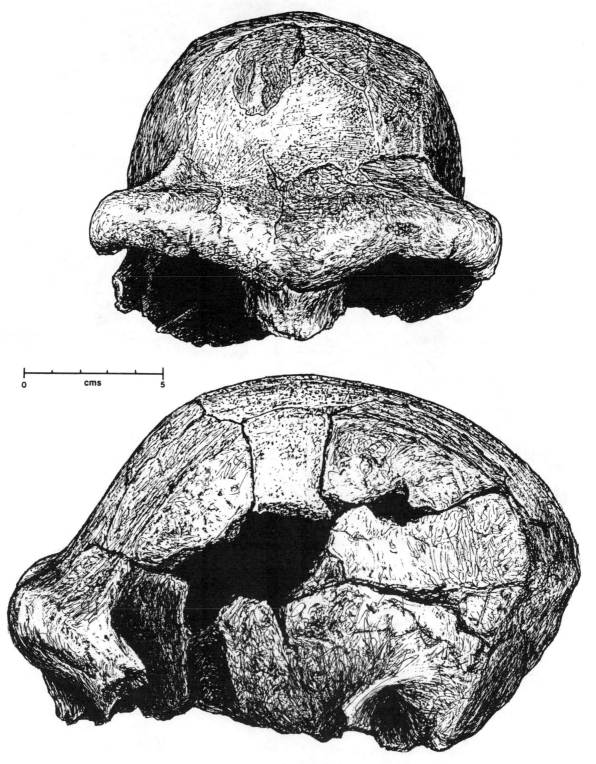

References: L. Leakey, 1961; Tobias, 1965a; M. D. Leakey, 1971; Holloway, 1973; Rightmire, 1979, 1981, 1990; Maier and Nkini, 1984, 1985; Wolpoff, 1998.

Bodo

Specimen: **Bodo**

Geographic location: Middle Awash Valley, Ethiopia

Taxonomic affiliation: *Homo erectus*

Dating: Middle Pleistocene (600,000 yBP)

General description: Recovered by Asfaw, Whitehead, and Wood (Rift Valley Research Mission in Ethiopia) at the site of Bodo in 1976, the specimen was found in nearly 100 fragments scattered about a wide area. It consists of most of the face. It has very archaic features that in a number of ways are similar to those of other *Homo erectus* in Africa. The frontal shows a general flattening, and the browridges are large. T. White, D. Clark, B. Asfaw, and G. Suwa have documented a series of cutmarks on the zygoma (see superior left zygoma below), orbit, nasal aperture, frontal, lacrimal, and parietals. White suggests that these cutmarks indicate that the face of this individual had been defleshed with a stone tool by another hominid prior to its deposition. This pattern has also been identified in Neandertal samples from Abri Moula (France) and may represent cannibalism.

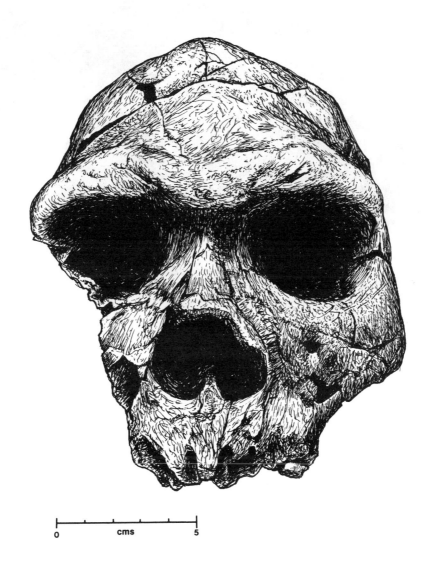

0   cms   5

References: Conroy, Jolly, Cramer, and Kalb, 1978; Kalb, Jolly, Oswald, and Whitehead, 1984; Asfaw, 1984; Rightmire, 1984, 1996; White, 1985, 1986; Defleur, Dutour, Valladas, and Vandermeersch, 1993; Clark, de Heinzelin, Schick, and others, 1994.

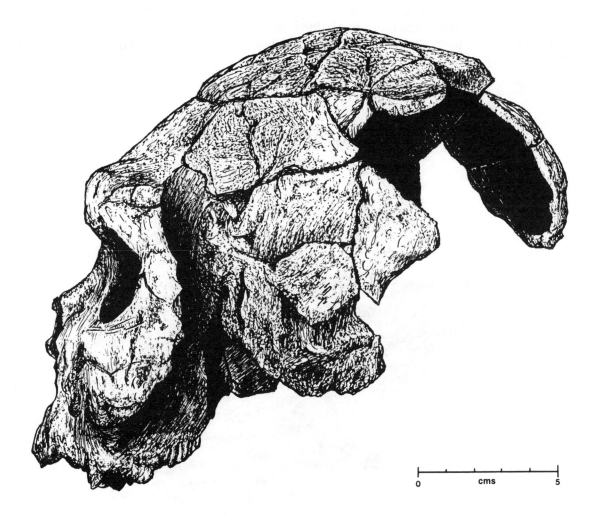

0   cms   5

Ternifine 2 and 3

Specimen: **Ternifine 2 and 3**
Geographic location: Ternifine, Algeria

Taxonomic affiliation: *Homo erectus*
Dating: Middle Pleistocene (700,000 yBP)

General description: Shown are two of three mandibles recovered from a sand pit in 1954 by C. Arambourg and R. Hoffstetter for the Musée National d'Histoire Naturelle. Although a number of paleoanthropologists have emphasized the differences within this sample of hominids, the differences most likely reflect normal population variability. Ternifine 2 (bottom) is comprised of a left half of a mandible and partial dentition; Ternifine 3 (top) is a nearly complete mandible and dentition. Both jaws are quite robust, with low and broad rami. They are similar in overall morphology to those of *Homo erectus* in China and Indonesia.

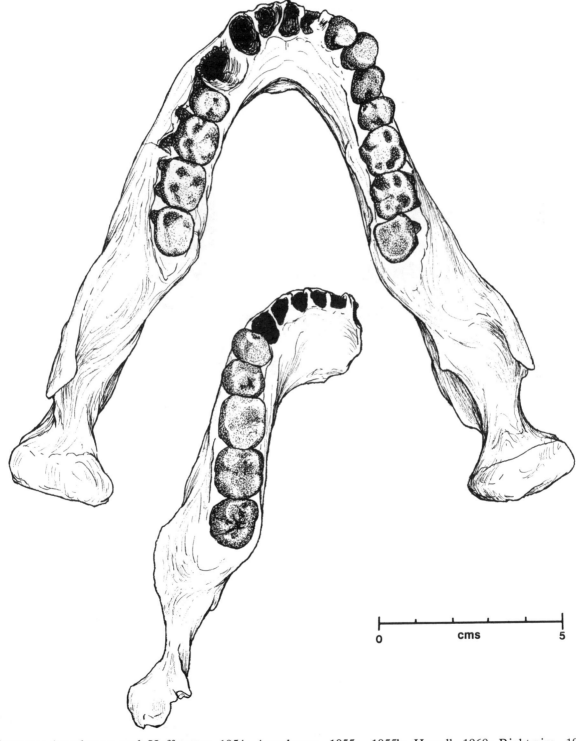

References: Arambourg and Hoffstetter, 1954; Arambourg, 1955a, 1955b; Howell, 1960; Rightmire, 1990; Wolpoff, 1998.

Mauer

Specimen: **Mauer**
Geographic location: Mauer, Germany

Taxonomic affiliation: *Homo erectus*
Dating: Middle Pleistocene (500,000 yBP)

General description: This fossil was discovered by workmen in a sand pit in 1907 and was studied by O. Schoetensack from the University of Heidelberg shortly thereafter. It consists of a complete mandible and most of the dentition. This mandible is quite large and robust, but in comparison with many other *Homo erectus* mandibles, it is small.

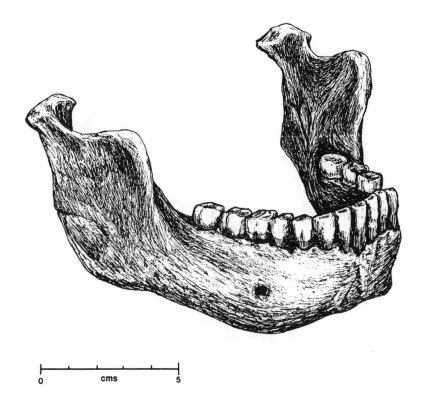

0   cms   5

References: Schoetensack, 1908; Hrdlička, 1930; Howell, 1960; Brace and Montagu, 1977; Stringer, Hublin, and Vandermeersch, 1984; Kraatz, 1985a, 1985b; Rightmire, 1990; Wolpoff, 1998.

Sangiran 4

Specimen: **Sangiran 4 (Pithecanthropus Skull IV)** Taxonomic affiliation: *Homo erectus*
Geographic location: Sangiran, Indonesia   Dating: Early Pleistocene (<1 myBP)

General description: Discovered in 1939 and first reported by R. von Koenigswald and F. Weidenreich, the original specimen consisted of the back three-quarters of the cranial vault and a partial maxilla with the dentition. The reconstruction shown here includes a mandible fragment from another individual that had been recovered in 1937. Sangiran 4 shows a number of important *Homo erectus* features—thickened bones of the cranial vault, and thickened and backward projecting attachment area for the neck muscles. The palate possesses a number of primitive characteristics, such as relatively large canines and a diastema between the canine and incisor. However, more modern features—reduced facial projection and size of the face in general—align it with the represented taxon, *Homo erectus*.

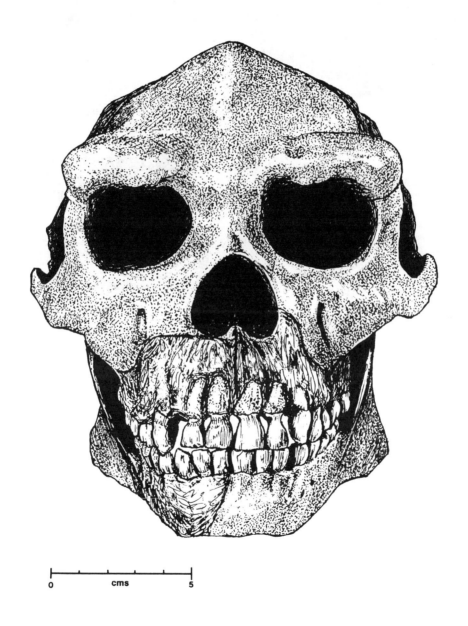

0   cms   5

References: von Koenigswald and Weidenreich, 1939; Weidenreich, 1940, 1945, 1946a, 1946b; Jacob, 1975a, 1975b; Pope, 1988; Rightmire, 1990; Wolpoff, 1998.

Sangiran 4

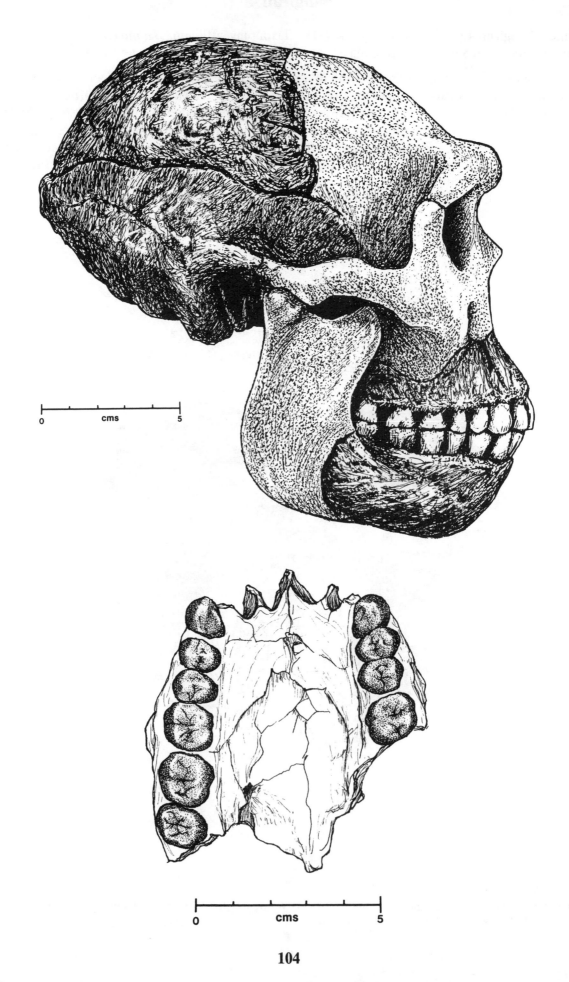

0 cms 5

0 cms 5

Sangiran 17

Specimen: **Sangiran 17 (Pithecanthropus Skull VIII)**   Taxonomic affiliation: *Homo erectus*

Geographic location: Sangiran, Indonesia   Dating: Middle Pleistocene (700,000 yBP)

General description: This distorted cranium—the most complete H. erectus from the site—was recovered by collectors for S. Sartono in 1969. The reconstruction shown here was subsequently completed by M. H. Wolpoff. Some of the obvious features of the cranium include a projecting face, long, broad and low overall cranial form, continuous browridges across the midline, and some postorbital constriction. In this regard, it is like other *Homo erectus* crania—especially that of Indonesian *Homo erectus* as well as the Chinese and East African variants.

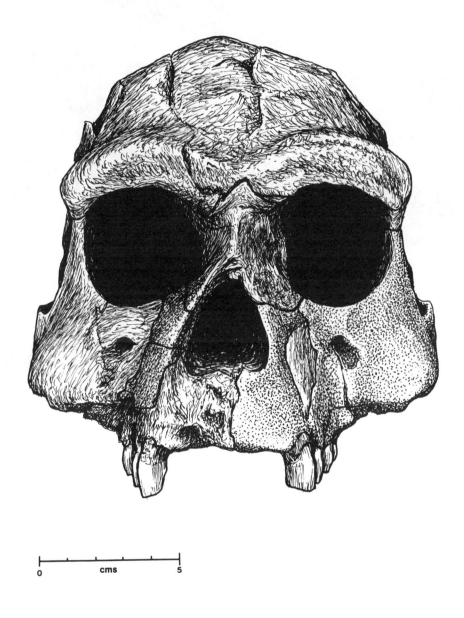

0   cms   5

References: Sartono, 1971, 1972, 1975; Jacob, 1975a, 1975b; Ninkovich and Burckle, 1978; Santa Luca, 1980; Thorne and Wolpoff, 1981; Holloway, 1981a; Sartono and Grimaud-Hervé, 1983; Pope, 1988; Rightmire, 1990; Wolpoff, 1998.

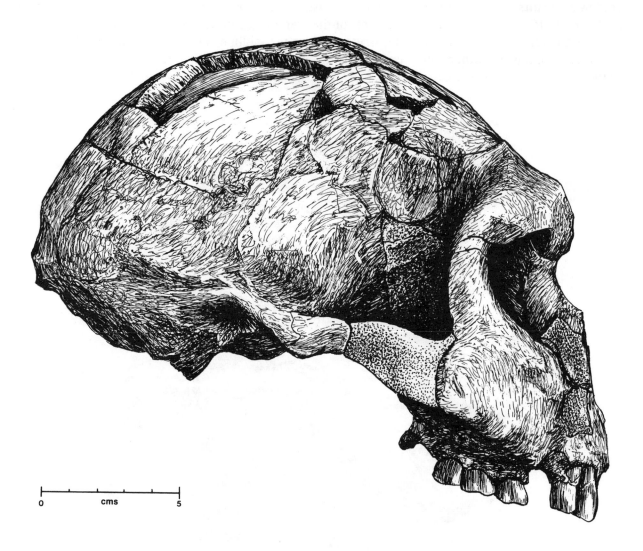

0   cms   5

Sangiran IX

Specimen: **Sangiran IX**
Geographic location: Sangiran, Indonesia

Taxonomic affiliation: *Homo erectus*
Dating: Middle Pleistocene (700,000 yBP)

General description: This fragmentary, but relatively undistorted, cranium was found by a local farmer near the village of Sangiran in 1993. Donald Tyler, an American physical anthropologist working in the area, was immediately contacted. In collaboration with S. Sartono and G. S. Krantz, he immediately provided a preliminary description of this important fossil. The cranium is similar to other *Homo erectus* individuals, including presence of a sagittal keel, gable-shape in posterior view, and very prominent supraorbitals with a well-developed sulcus above them. Preliminary assessment of cranial capacity is at 856 cc.

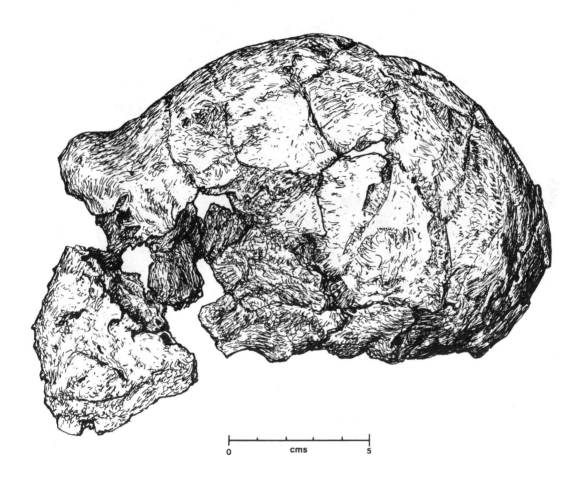

References: Eijjgenraam, 1993; Tyler, Sartono, and Krantz, 1994; Tyler, personal communication; Wolpoff, 1996.

Hexian

Specimen: **PA 830, Hexian**

Geographic Location: Longtandong Cave, Hexian County, Anhui Province, People's Republic of China

Taxonomic affiliation: *Homo erectus*

Dating: Middle Pleistocene (150,000–270,000 yBP)

General description: Found in 1980, this specimen consists of a cranial vault. In general, this individual shows typical *Homo erectus* features—flattened, receding forehead, well developed browridges, thick cranial vault bones, and a long, low front-to-back profile. In overall details, this cranium is similar to those of *Homo erectus* hominids from both Zhoukoudian and Indonesia. The cranium appears more modern than that of *Homo erectus* from Zhoukoudian in that it shows a reduction in postorbital constriction.

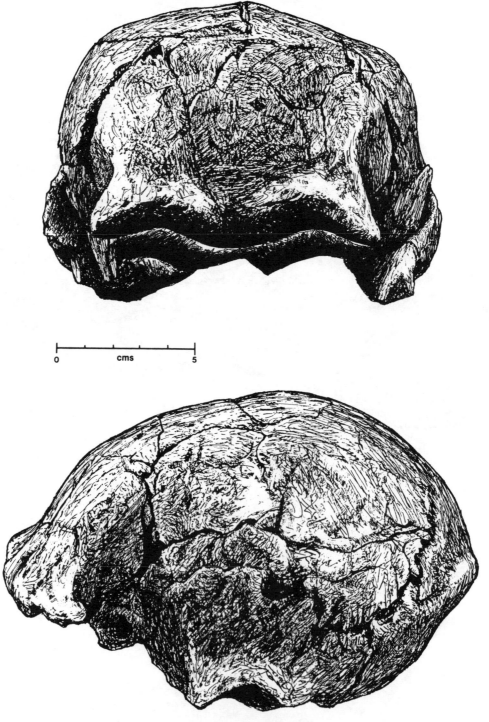

References: Wu Rukang and Dong Xingren, 1982, 1983, 1985; Wu Rukang, 1983; Wolpoff, Wu and Thorne, 1984; Rightmire, 1990; Wu and Poirier, 1995; Etler, 1996.

Zhoukoudian

Specimen: Zhoukoudian skull reconstruction
Geographic location: Locality 1, Zhoukoudian (formerly Choukoutien), People's Republic of China

Taxonomic affiliation: *Homo erectus*
Dating: Middle Pleistocene (300,000–550,000 yBP)

General description: The composite reconstruction by Franz Weidenreich was based on cranial materials recovered from the site prior to World War II. The reconstruction of the face is conjectural because few facial bones were found at the site. The later finding of *Homo erectus* crania with faces at other sites, however, suggests that the reconstruction is essentially accurate. The cranium is long and low with thick cranial bones. The forehead is higher and more filled out relative to those of earlier hominids.

The site is of tremendous importance in paleoanthropology, primarily because a large collection of human remains, faunal remains, and tools, as well as other evidence of prehistoric occupation that was associated with *Homo erectus* were recovered from a series of well-dated stratigraphic layers. With the exception of materials collected from Zhoukoudian by Chinese paleontologists and archaeologists after World War II, all other remains disappeared in 1941 at the outbreak of World War II. Fortunately, while still in Beijing, Franz Weidenreich made complete sets of cast reproductions of the hominid materials before their disappearance. These casts, in addition to a series of detailed reports (see References), has made it possible to continue study of this valuable fossil sample.

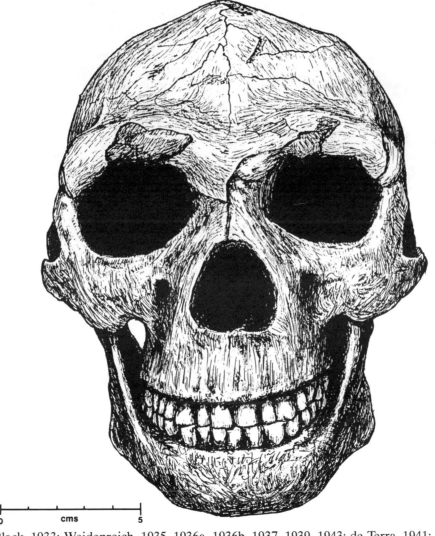

0   cms   5

References: Black, 1933; Weidenreich, 1935, 1936a, 1936b, 1937, 1939, 1943; de Terra, 1941; Chang, 1962; Chiu Chung-lang, Ku Yu-min, Chang Yin-yun, and Chang Sen-shui, 1973; Shapiro, 1974; Howells, 1977; Institute of Vertebrate Paleontology and Paleoanthropology (Chinese Academy of Sciences), 1980; Jia Lanpo, 1980; Mann, 1981; Trinkaus, 1982b; Wu Rukang, 1985; Wu Rukang and Dong Xingren, 1983, 1985; Wu Rukang and Lin Shenglong, 1983; Liu Ze-Chun, 1983; Pope, 1988; Brooks and Wood 1990; Rightmire, 1990; Wu and Poirier, 1995; Tattersall and Sawyer, 1996; Wolpoff, 1996; Etler, 1996; Grün, 1997.

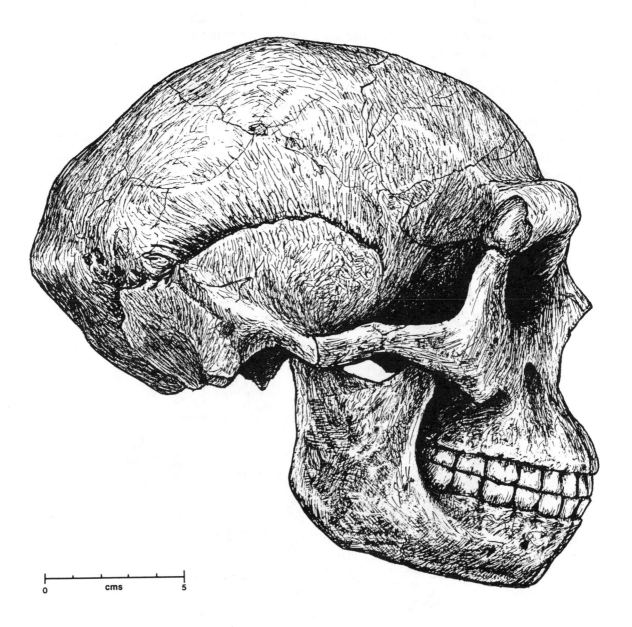

0   cms   5

Early Archaic *Homo sapiens*

Broken Hill 1

6. EARLY ARCHAIC *HOMO SAPIENS*

Middle Pleistocene Hominids

This section illustrates a series of hominids that are representative of the earliest of our own species, *Homo sapiens.* The boundary between these and earlier hominids is arbitrary. However, the period of time covers from somewhat less than 400,000 years ago to about 100,000 years ago. Although still quite primitive in a number of respects (e.g., thick and robust cranial bones), the reduction in size of facial bones and cheek teeth as well as the expansion in size of the front teeth and endocranial volume sets this group apart from their *Homo erectus* forebears. In comparison with *Homo erectus*, there is good evidence for morphological continuity. For example, specimens of early archaic *Homo sapiens* from Asia share a number of similarities with late *Homo erectus* from the same continent. The same case can be made for Africa and Europe, although the number of fossil samples from these continents is more limited than in Asia.

The fossils that are represented in this sample of Hominidae include the following:
1. Kabwe, Zambia: Broken Hill 1
2. Lake Ndutu, Tanzania: Ndutu
3. El Hamra, Morocco: Salé
4. Petralona, Greece
5. Tautavel, France: Arago 21
6. Sierra de Atapuerca, Spain: Atapuerca 5
7. Swanscombe, England
8. Steinheim, Germany
9. Madhya Pradesh, India: Narmada
10. Ngandong, Indonesia: Ngandong 7
11. Shaanxi, China: Dali

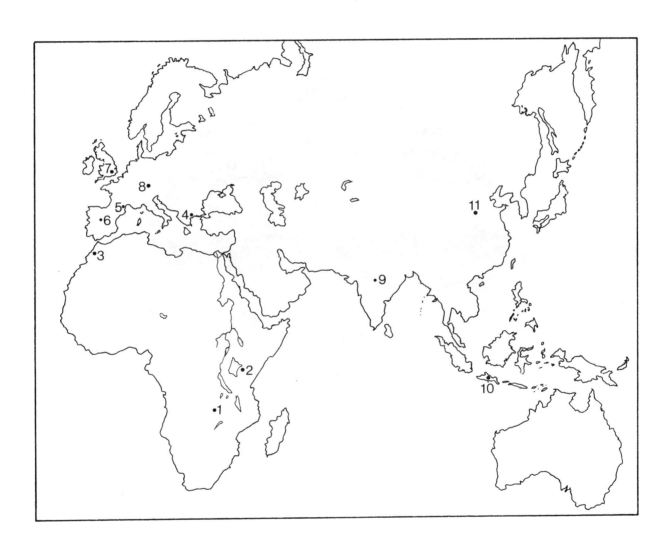

Broken Hill 1

Specimen: **Broken Hill 1 (BMNH 686)**
Geographic location: Broken Hill Mine,
  Kabwe, Zambia

Taxonomic affiliation: *Homo sapiens*
Dating: Middle Pleistocene (125,000 yBP)

General description: This specimen is one of several individuals recovered in 1921 during opencast mining in a limestone hill. The dating of the cranium is uncertain, but comparisons of associated fauna with similar forms from well-dated south and east African sites suggest a middle Pleistocene context. The cranium shows a combination of features that are both rugged and gracile. The browridges and occipital are very well developed, but the expansion of the cranium, reduced attachment sites for neck musculature, reduced size of facial bones, and thinner cranial vault bones indicate an association with archaic *Homo sapiens*.

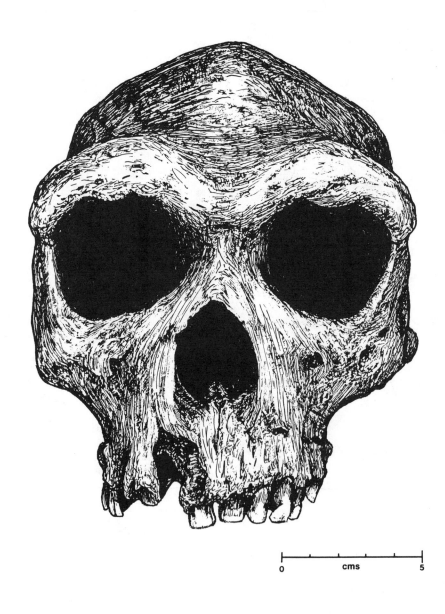

0   cms   5

References: Woodward, 1921; Pycraft and others, 1928; Le Gros Clark, 1928; Hrdlička, 1930; Clark, Oakley, Wells, and McClelland, 1947; Klein, 1973; Rightmire, 1976; Koritzer and St. Hoyme, 1977, 1980; Howell, 1978; Murrill, 1981; Skinner and Sperber, 1982; Bartsiokas and Day, 1993; Wolpoff, 1998.

Ndutu

Specimen: **Ndutu**

Geographic location: Lake Ndutu, Tanzania

Taxonomic affiliation: *Homo sapiens*

Dating: Middle Pleistocene (350,000 yBP)

General description: This badly fragmented and incomplete cranium was found by A. A. Mturi (Tanzania Department of Antiquities) in 1973 and subsequently reconstructed by R. J. Clarke. The cranium represents an early archaic form of *Homo sapiens*, that, like the Broken Hill cranium, exhibits a series of primitive features reminiscent of *Homo erectus* (e.g., thickened occipital, thick cranial vault bones). However, the expanded cranium and general contour of the cranium align this specimen with the archaic *Homo sapiens*.

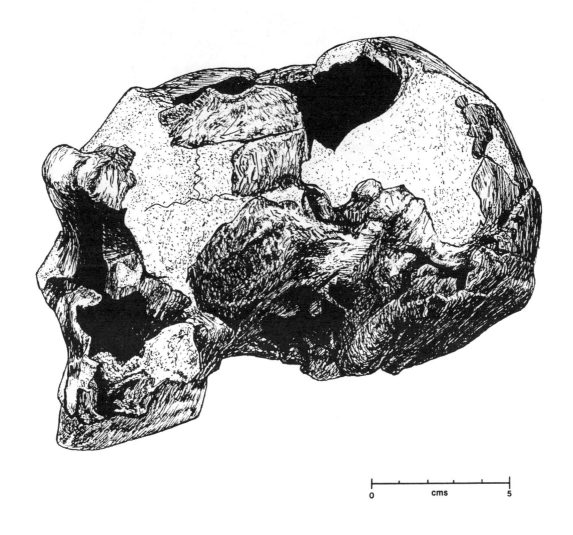

0   cms   5

References: Mturi, 1976; Clarke, 1976, 1990; Rightmire, 1983, 1984; Wolpoff, 1998.

Salé

Specimen: **Salé**

Geographic location: El Hamra, Morocco

Taxonomic affiliation: *Homo sapiens*

Dating: Middle Pleistocene (200,000–250,000 yBP)

General description: Discovered by quarrymen and collected by J. Jaeger in 1971, this well-preserved north African hominid shows a number of similarities with sub-Saharan early archaic *Homo sapiens*. Most of the face is missing in the probable female, but it is possible to see distinct postorbital constriction. Although recent estimates of endocranial capacity are quite small (880 cc), the more filled-out appearance of the cranial vault—especially the sides—is suggestive of an endocranial expansion relative to earlier grades of hominid evolution. The area of the attachment for the neck muscles on the occipital bone is very small, which may be pathological and not normal.

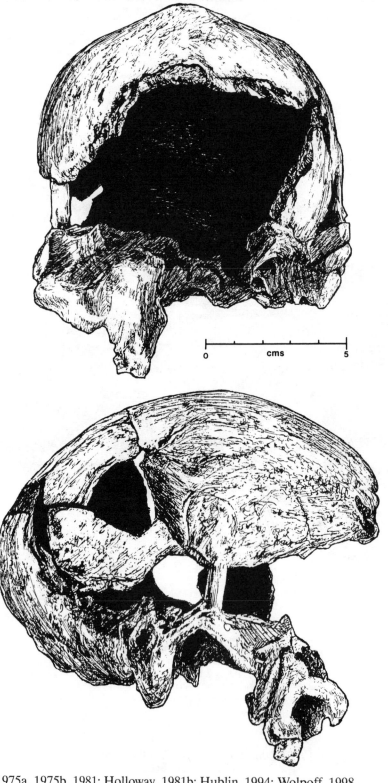

References: Jaeger, 1975a, 1975b, 1981; Holloway, 1981b; Hublin, 1994; Wolpoff, 1998.

Petralona

Specimen: **Petralona**
Geographic location: Petralona, Greece

Taxonomic affiliation: *Homo sapiens*
Dating: Middle Pleistocene (150,000–250,000 yBP)

General description: This specimen is one of the best preserved crania from the middle Pleistocene in Europe. Discovered in a cave by local villagers in 1960, the context of the specimen within the site has remained clouded. Due to the uncertain context of the cranium, dating is problematical. The cranium exhibits a rather large face with a number of *erectus*-like features, including low endocranial capacity (1200 cc) and thick cranial vault bones. However, the presence of a relatively greater height of the cranium and overall expansion as well as a lightening of cranial superstructure align it with early archaic *Homo sapiens*. The morphology of the face foreshadows features seen in later archaic *Homo sapiens*.

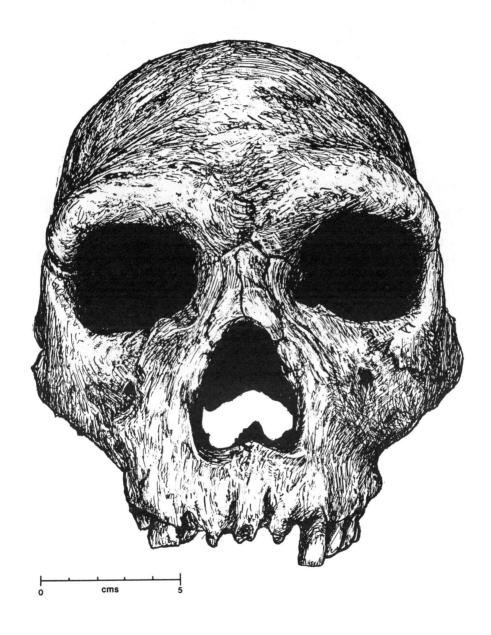

0   cms   5

References: Poulianos, 1971, 1982, 1989; Stringer, 1974; Smith, 1977; Stringer, Howell, and Melentis, 1979; Gantt, Xirotiris, Kurten, and Melentis, 1980; Hennig, Herr, Weber, and Xirotiris, 1981; Murrill, 1981; Liritzis, 1982; Ikeya, 1982; Grün, 1996.

117

0 cms 5

Arago 21

Specimen: **Arago 21**
Geographic location: Tautavel, France

Taxonomic affiliation: *Homo sapiens*
Dating: Middle Pleistocene (200,000–400,000 yBP)

General description: One of 23 individuals recovered from a cave in southeastern France by Henry and Marie-Antoinette de Lumley in the late 1960s and early 1970s, this specimen is represented by a slightly deformed, yet complete, face. The browridges are prominent and large. The cranium combines a small face with large zygomas. The forehead is flattened. This specimen is a good representative of the transition from *Homo erectus* to *Homo sapiens* in Europe.

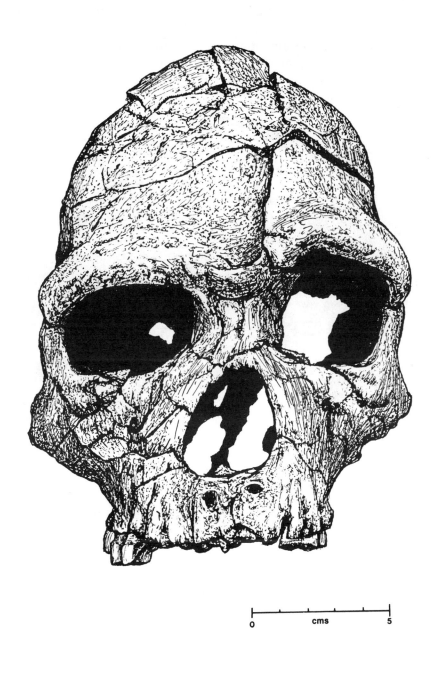

0   cms   5

References: H. de Lumley and M.-A. de Lumley, 1973; M.-A. de Lumley, 1981; Stringer, Hublin, and Vandermeersch, 1984; Wolpoff, 1998.

Atapuerca 5

Specimen: **Atapuerca 5**
Geographic location: Sierra de Atapuerca, Burgos, Spain

Taxonomic affiliation: *Homo sapiens*
Dating: Middle Pleistocene (>300,000 yBP)

General description: This relatively complete cranium—the most complete premodern cranium known—is one of some 1600 hominid bones representing at least 32 individuals recovered from a limestone cave in northern Spain called Sima de los Huesos ("Pit of Bones"). This amazing sample represents all regions of the human skeleton ranging in age from 4 to 35 years, and is is the largest Middle Pleistocene collection of human fossils from a single locality. The cranium shown here is one of three important fossils discovered by Juan Luis Arsuaga of Complutense University, Madrid, in the summer of 1992. Reminiscent of *Homo erectus* in some respects, the individual shows a well-defined, large supraorbital torus. On the whole, this and other crania from the site display a suite of characteristics that are consistent with a pattern of local evolution, with a number of traits (e.g., pronounced midfacial projection) foreshadowing the appearance of Neandertals. The cranial capacity of this individual is 1125 cc.

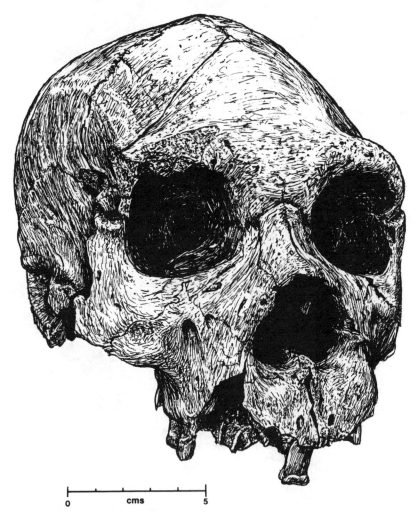

0   cms   5

References: Arsuaga, Martínez, Cracia, Carretero, and Carbonell, 1993; Stringer, 1993; Pares and Perez-Gonzalez, 1995; Johanson and Blake, 1996; Bahn, 1996; Wolpoff, 1996, 1998; Bermúdez de Castro, Arsuaga, Carbonell, Rosas, Martínez, and Mosquera, 1997; Arsuaga, Bermúdez, and Carbonell, 1997; Arsuaga, Martínez, Gracia, Carretero, Lorenzo, and García, 1997; Bischoff, Fitzpatrick, León, Arsuaga, Flagueres, Bahain, and Bullen, 1997; Andrews and Fernandez Jalvo, 1997; Arsuaga, Martínez, Gracia, and Lorenzo, 1997; Martínez and Arsuaga, 1997; Rosas, 1997; Carretero, Arsuaga, and Lorenzo, 1997; Pérez, Gracia, Martínez, and Arsuaga, 1997.

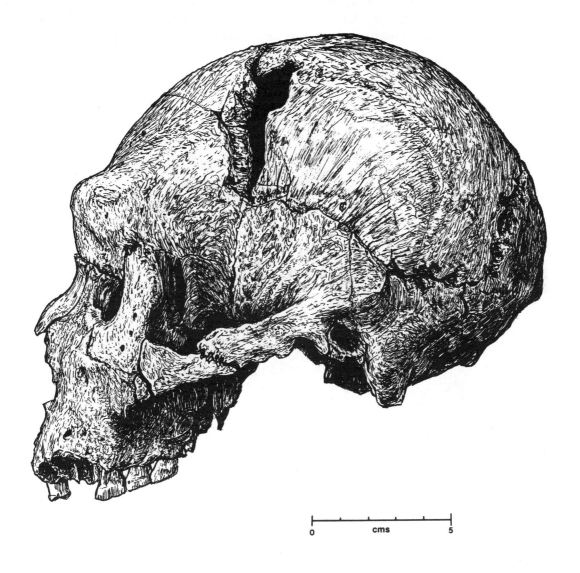

0   cms   5

Swanscombe

Specimen: **Swanscombe**
Geographic location: Swanscombe, England

Taxonomic affiliation: *Homo sapiens*
Dating: Middle Pleistocene (200,000–250,000yBP)

General description: This fossil is represented by an occipital and left parietal that were recovered in the mid-1930s by A. T. Marston and a right parietal recovered in 1955 by J. Wymer and A. Gibson. Unfortunately, no face is present, but the overall morphology of materials at hand is suggestive of both a large area of neck muscle attachment and a relatively small endocranial capacity (1250 cc). The form of the cranium is rounded. The face was probably quite similar to the cranium from Steinheim.

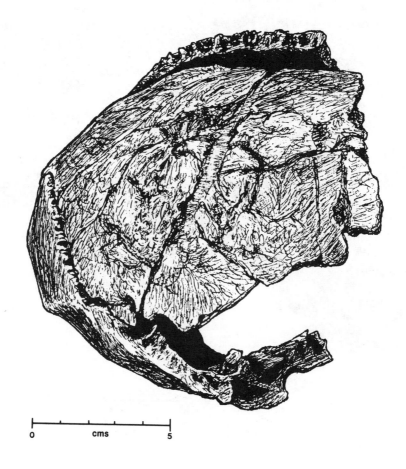

0   cms   5

References: Marston, 1936, 1937; Oakley, 1952; Wymer, 1955; Howell, 1960; Sergi, 1962; Weiner and Campbell, 1964; Ovey, 1964; Stringer, Hublin, and Vandermeersch, 1984; Conway, McNabb, and Ashton, 1996; Wolpoff, 1998.

Steinheim

Specimen: **Steinheim**

Geographic location: Steinheim an der Murr, Germany

Taxonomic affiliation: *Homo sapiens*

Dating: Middle Pleistocene (200,000–250,000 yBP)

General description: This cranium was discovered by K. Sigrist in a gravel pit in 1933. Upon discovery, it was removed from the burial matrix under the direction of the curator of the Staatliches Museum für Naturkunde (Stuttgart), Fritz Berckhemer. The Steinheim cranium is comprised of a nearly complete cranial vault and most of the face. The greater part of the left side of the face and occipital are crushed inward. This is a long, low cranium with a small vault, small mastoid processes, and a gracile face. Although the cranial vault has a somewhat filled out appearance, the endocranial capacity is relatively small for an archaic *Homo sapiens* (1100 cc) and falls well within the range of *Homo erectus* endocranial capacity. This specimen is an important transitional form between *Homo erectus* and *Homo sapiens*.

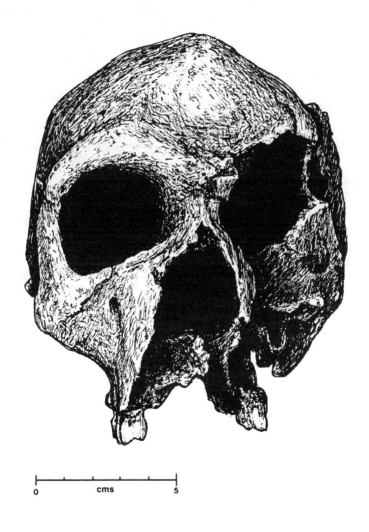

0   cms   5

References: Weinert, 1936; Howell, 1960; Stringer, Hublin, and Vandermeersch, 1984; Adam, 1985; Wolpoff, 1998.

123

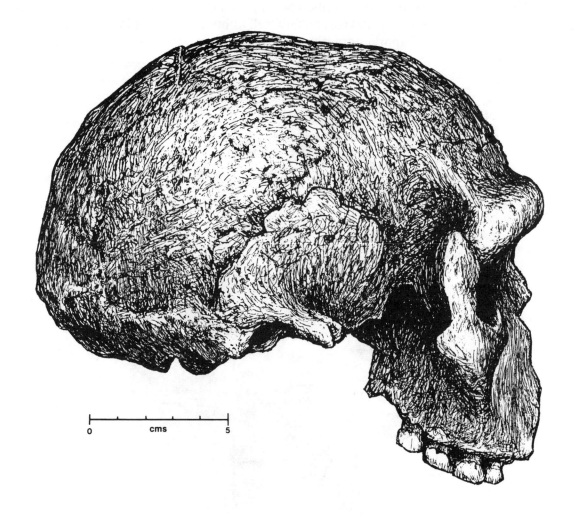

0   cms   5

Narmada

Specimen: **Narmada**
Geographic location: Hathnora, Madhya
  Pradesh, India

Taxonomic affiliation: *Homo sapiens*
Dating: Middle to Late-Middle Pleistocene
(<150,000 yBP)

General description: Found in 1982, this partial cranium is the only hominid fossil dating to the Middle Pleistocene from the Indian subcontinent. The cranium perserves the right frontal, sphenoid, zygoma, temporal, occipital, and left parietal; the face, mandible, and dentition are missing. The individual is likely female. The relative thickness of the cranial bones is similar to that of Neandertals, and the occipital curvature is like Asian archaic *Homo sapiens* (e.g., Dali, Ngandong 12). The cranial capacity is between 1155 and 1421 cc.

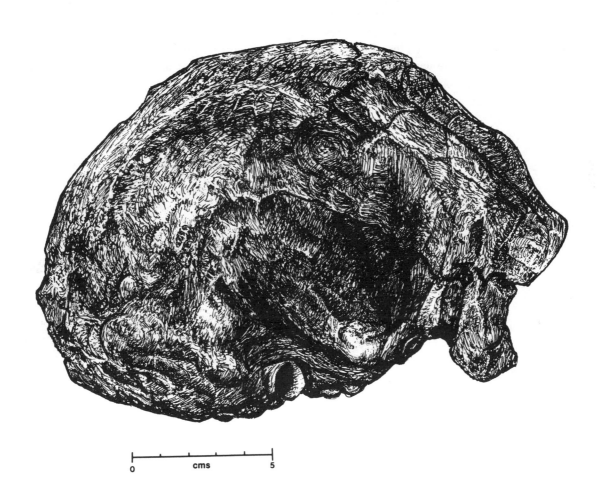

0   cms   5

References: Sonakia, 1985; Badam, 1989; Kennedy, Sonakia, Chiment, and Verma, 1991; Kennedy, 1994.

Ngandong 7

Specimen: **Ngandong 7 (Solo Skull VI)**
Geographic location: Ngandong, Indonesia

Taxonomic affiliation: *Homo sapiens*
Dating: Middle Pleistocene (250,000 yBP; 27,000–53,000 yBP)

General description: Shown here is one of eleven crania recovered during the period of 1931 to 1933 by C. ter Haar and G. H. R. von Koenigswald. Subsequent final preparation and study was completed by Franz Weidenreich. This specimen is the most complete of the series, but like the others, it is missing the face. Relative to *Homo erectus* from Indonesia, the cranium is large, a factor that reflects an expansion of the endocranial volume (average for the series = 1150 cc). Other notable changes in the vault in comparison with *Homo erectus* include greater height and expansion of the frontal and occipital regions. The browridge, moreover, is less pronounced, a development that is due, in part, to the expanded frontal. Some workers include the Ngandong sample with *Homo erectus* in their taxonomic assessments. However, the overall morphology of the crania in this series is indicative of an affiliation with *Homo sapiens*.

Electron spin resonance and uranium dates of 27,000–53,000 recently determined from fossil bovid teeth—collected at the site decades after the original excavations—suggest that early archaic *Homo sapiens* in Indonesia co-date with modern humans. However, the dates may be problematical because the co-occurrence of the fossil hominids and bovids is questionable; only direct dating of the fossil hominids will provide an accurate age.

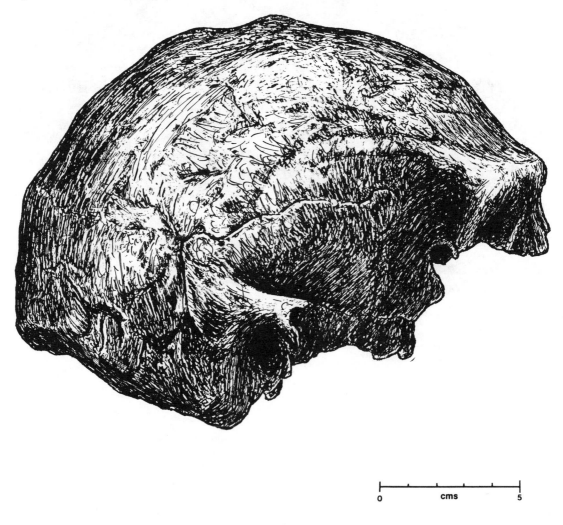

0   cms   5

References: Oppenoorth, 1937; Weidenreich, 1951; Swisher, Rink, Antón, Schwarcz, Curtis, Suprijo, and Widiasmoro, 1996; Gibbons, 1996; Grün and Thorne, 1997; Wolpoff, 1998.

Dali

Specimen: Dali

Geographic location: Dali County, Shaanxi Province, People's Republic of China

Taxonomic affiliation: *Homo sapiens*

Dating: Middle Pleistocene (209,000 yBP)

General description: This cranium was found in 1978 by Shuntang Liu and others under the auspices of various scientific agencies in the People's Republic of China, including the Institute of Vertebrate Pale-ontology and Paleoanthropology. This fossil hominid consists of a nearly complete cranium—only part of the left zygoma, maxilla, and parietal is missing. The robusticity of the individual suggests that it is a male. The endocranial capacity is 1120 cc, which is on the upper range for variation seen in *Homo erectus*, but on the lower range for *Homo sapiens*. A number of features are *erectus*-like, including thick cranial bones, large supraorbital torus, low cranial vault, and the presence of a keel running along the midline of the top of the cranium. However, the overall construction of the cranium—length, breadth, height, vertical orientation of the sides of the cranium, a greater degree of filling out—are suggestive of a more advanced form of hominid, almost certainly archaic *Homo sapiens*.

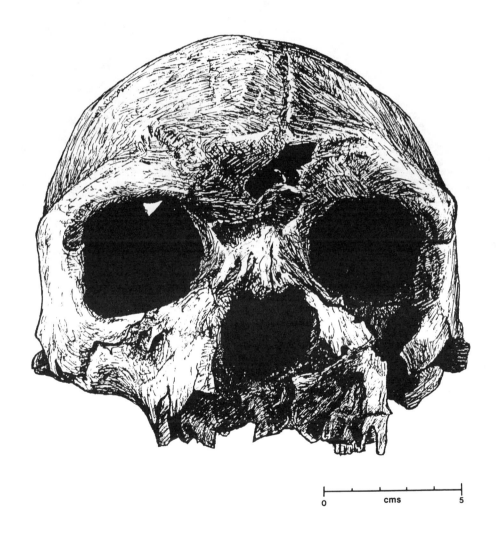

References: Wu Xinzhi, 1981; Wu Xinzhi and Wu Maolin, 1985; Wu Rukang, 1983; Brace, Shao Siang-qing, and Zhang Zhen-biao, 1984; Pope, 1988; Smith, Falsetti, and Donnelly, 1989.

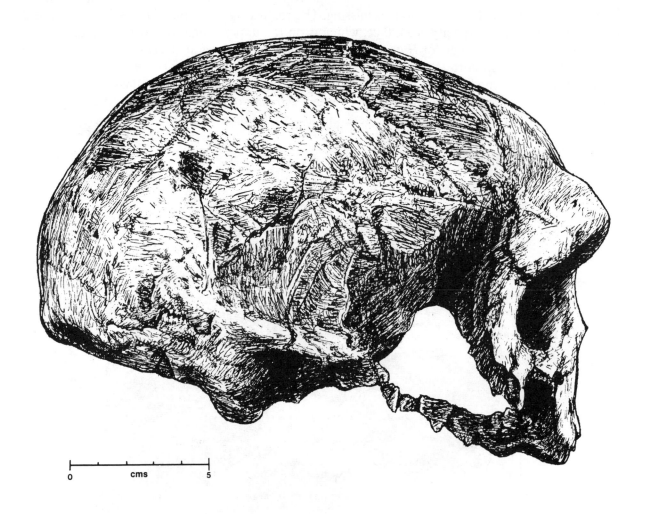

0   cms   5

Late Archaic *Homo sapiens*

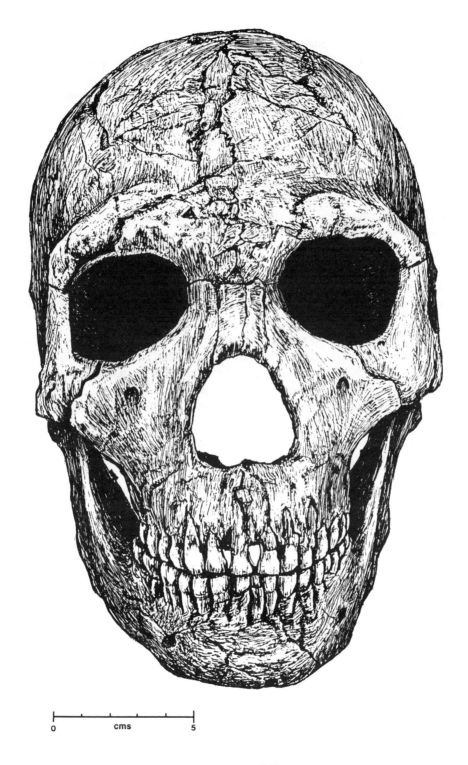

Amud I

7. LATE ARCHAIC *HOMO SAPIENS*

Upper Pleistocene Hominids

Populations of hominids that are representative of this period of human evolution include the Neandertals of Europe and the Near East and select contemporary populations in the remainder of Asia and Africa. Paleoanthropological research on this period of human evolution has seen a resurgence in recent years. Many new questions, new fossils, and new data are currently under investigation by a variety of scientists (Trinkaus, 1986; Smith, Falsetti, and Donnelly, 1989). An important question that is being addressed by paleoanthropologists is the extent to which the transition from anatomically archaic populations to anatomically modern populations is represented by population dispersal and subsequent replacement or by gradual, local or regional evolutionary change. This time period spans roughly 100,000 years ago to 30,000 years ago when anatomically modern humans are present in all areas of the Old World.

In comparison with earlier hominids, we see increased brain size as well as a number of distinctive craniofacial changes. The earlier sample of teeth (e.g., Krapina) exhibits an expansion in size of the anterior dentition and reduction in size of the posterior dentition. Later hominids from this sample show reductions in size of the teeth in both the front and back of the mouth. Despite reduction in the size of the face and jaws, there is a clear forward projection of the midface, especially in European Neandertals. In addition, the nose is large and protruding, and in at least some individuals, the foramina below the eye orbits and the maxillary sinuses are large. These characteristics represent probable adaptations to cold, dry conditions of the Upper Pleistocene. The crania of hominids during this period show a characteristic flattening and elongation of the upper portion of the occipital called bunning. Although specific mechanical factors may be involved in providing for this morphology (Wolpoff, 1998), Trinkaus and LeMay (1982) have suggested that bunning may have arisen from differential growth of the back portion of the brain prior to final growth of posterior cranial bones. The pressure exerted by the growing brain on the cranium may have resulted in the posterior displacement of the occipital. The hominids possessed muscular and robust bodies with especially strong arms and hands.

The fossil remains from this grade of human evolution include the following:
1. Laetoli, Tanzania: L.H. 18
2. Omo, Ethiopia: Omo 2
3. Jebel Irhoud, Morocco: Jebel Irhoud I
4. Kurdistan, Iraq: Shanidar I
5. Wadi Amud, Israel: Amud I
6. Wadi el-Mughara, Israel: Tabun Cl
7. Forbes' Quarry, Gibraltar: Gibraltar 1
8. Krapina, Croatia: Krapina 3
9. La Chapelle-aux-Saints, France
10. Dordogne, France: La Ferrassie 1
11. Charente, France: La Quina 5 and 18
12. Latina, Italy: Mt. Circeo 1
13. Rome, Italy: Saccopastore 1
14. Hochdahl, Germany: Neandertal

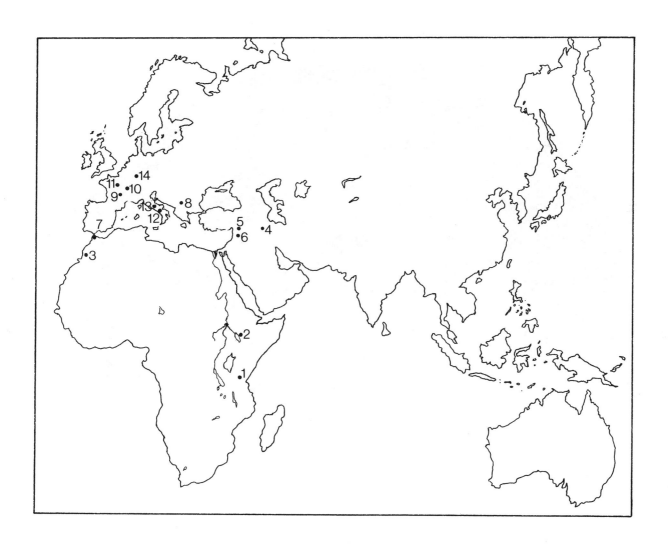

L.H. 18

Specimen: **L.H. 18**
Geographic location: Laetoli, Tanzania

Taxonomic affiliation: *Homo sapiens*
Dating: Upper Pleistocene (120,000 yBP)

General description: This nearly complete cranium was recovered from the Ngaloba Beds at Laetoli in 1976 by E. Kandini under the direction of Mary Leakey. The morphology of this individual is represented by a mixture of both modern and archaic features. Modern features include an expanded vault, gracile face, and rounded occipital; archaic features are evident from the low vault, flattened frontal, large browridges, and thick, robust cranial vault bones. This cranium represents an important transitional link between earlier and later forms of *Homo sapiens* in Africa.

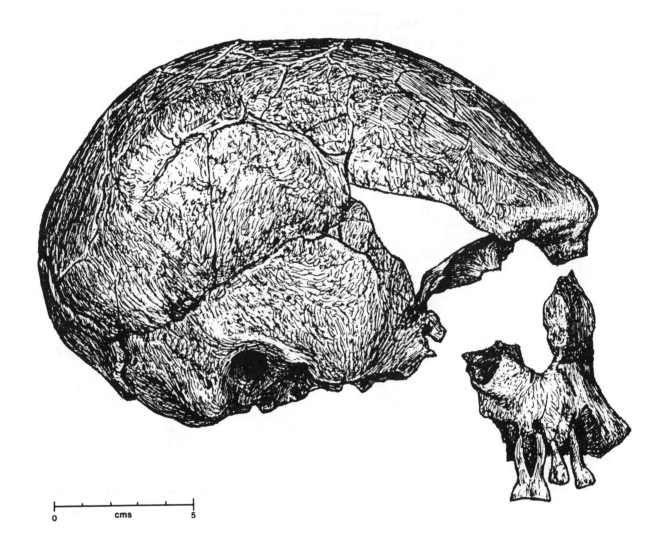

0 cms 5

References: Day, M. D. Leakey, and Magori, 1980; Magori, 1980; Magori and Day, 1983; Rightmire, 1984; Brauer, 1989; Smith, Falsetti, and Donnelly, 1989; Cohen, 1996; Wolpoff, 1998.

Omo 2

Specimen: **Omo 2**

Geographic location: Omo, Ethiopia

Taxonomic affiliation: *Homo sapiens*

Dating: Upper Pleistocene (60,000 yBP)

General description: Recovered by the Kenya contingent of the Omo Research Expedition under the direction of Richard Leakey in 1967, this cranial vault is one of three discovered from upper Pleistocene deposits at Omo. The cranium was a surface find, so dating is very imprecise. This specimen—lacking the face—is the most complete of the three. The cranium is long and low, and the frontal is flat. The curved nature of the posterior vault and the large area for neck muscle attachment are *erectus*-like, but the large endocranial volume (1400 cc) is a marker of a more advanced grade of evolution. The supraorbital is modern in appearance in that it is only a slight protrusion.

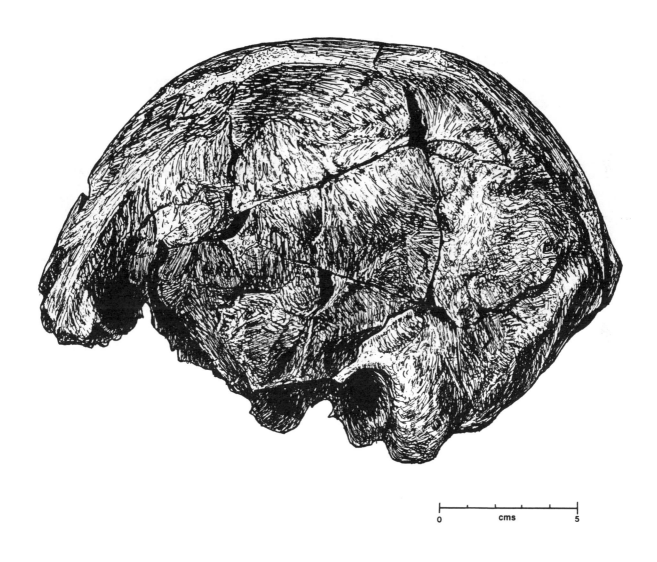

References: R. Leakey, 1969; Day, 1969; Rightmire, 1984; Wolpoff, 1998.

Jebel Irhoud 1

Specimen: **Jebel Irhoud 1**
Geographic location: 60 km. southeast
    of Safi, Morocco

Taxonomic affiliation: *Homo sapiens*
Dating: Upper Pleistocene (47,000 yBP)

General description: This cranium was discovered by a miner and, shortly thereafter, studied by E. Ennouchi in the early 1960s. The specimen is a relatively complete cranium, but it is missing the dentition. Morphologically, this individual is very similar to western European Neandertals in that the vault is long and low, the occipital shows a distinct backward projection, the forehead slopes, and the brow-ridges are distinct. The face, however, possesses a number of modern features, including especially a lack of pronounced projection of the midface.

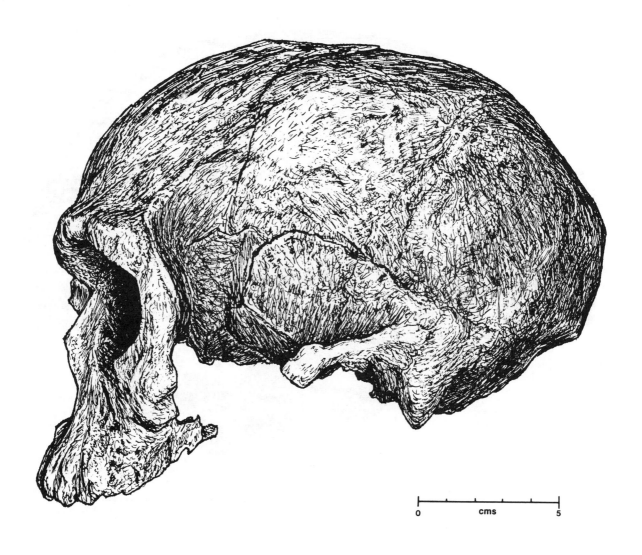

References: Ennouchi, 1962, 1963; Mann and Trinkaus, 1973; Jaeger, 1981; Smith, Falsetti, and Donnelly, 1989; Wolpoff, 1998.

Shanidar 1

Specimen: **Shanidar 1**
Geographic location: Kurdistan, Iraq

Taxonomic affiliation: *Homo sapiens*
Dating: Upper Pleistocene (46,000 yBP)

General description: This individual is represented by most of a skeleton recovered from a cave burial in 1957 by R. Solecki of Columbia University. Shown here is the nearly complete skull with associated dentition restored by T. D. Stewart. It exhibits features similar to other Neandertals, particularly with respect to being long and narrow, and exhibiting marked projection of the nasal area and midfacial region. Unlike other Neandertals, however, the frontal bone is quite straight, so much so that Trinkaus has suggested that the individual may have experienced artificial cranial deformation as an infant. The dentition, especially the anterior dentition, is extremely worn. The front teeth possess a rounded wear that may reflect their use in extra-masticatory functions. In addition to a number of postcranial fractures and other trauma-related injuries (see Trinkaus and Zimmerman, 1982), there are several traces of past injuries that are evident in the cranium. These include a crushing fracture to the left side of the eye orbit that could very well have caused blindness in the left eye as well as other disabilities. Additionally, some damage to the outer surface of the cranial vault (frontal bone) is suggestive of a scalp wound that did not involve fracture of the bone.

0   cms   5

Shanidar 1

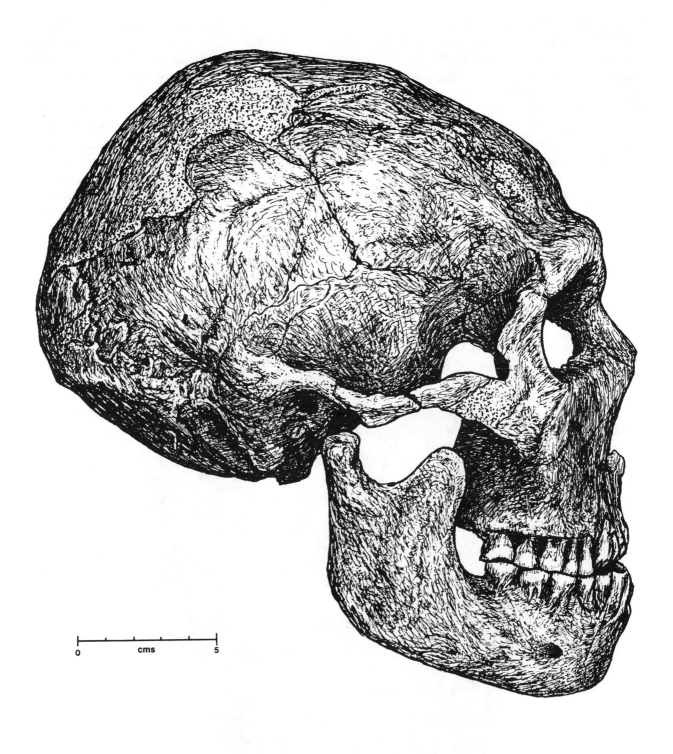

0   cms   5

References: Solecki, 1957, 1963, 1971; Stewart, 1958, 1961, 1977; Trinkaus, 1978, 1982a, 1983, 1984; Trinkaus and Zimmerman, 1982; Wolpoff, 1998.

Amud 1

Specimen: **Amud 1**

Geographic location: Wadi Amud, Israel

Taxonomic affiliation: *Homo sapiens*

Dating: Upper Pleistocene (35,000–45,000 yBP)

General description: Recovered in 1961 by a paleontological team from Tokyo University under the direction of H. Suzuki, this individual consists of a nearly complete skeleton with the cranium and mandible. Most of the face is missing, but was reconstructed by Suzuki following Shanidar 1 as a model. The cranium shows features that characterize other Neandertals and has the largest endocranial capacity of the Near Eastern Neandertals (1750 cc).

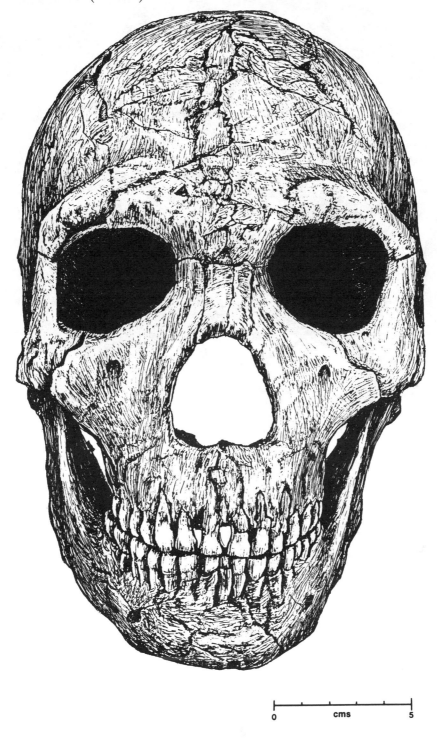

0   cms   5

References: Suzuki and Takai, 1970; Grün and Stringer, 1991; Wolpoff, 1998.

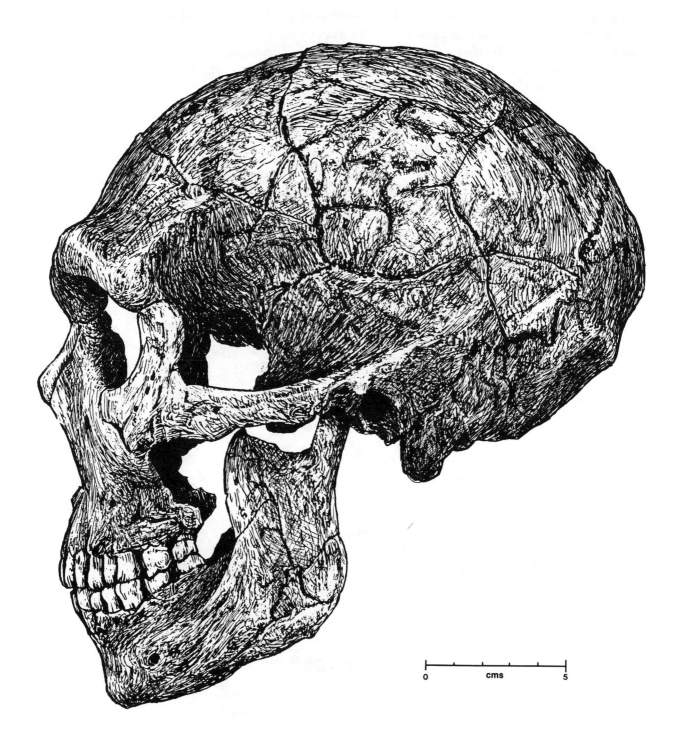

0   cms   5

Tabun C1

Specimen: **Tabun C1**

Geographic location: Wadi el-Mughara, Israel

Taxonomic affiliation: *Homo sapiens*

Dating: Upper Pleistocene (50,000 yBP; 85,000–130,000 yBP)

General description: Recovered by the Joint Expedition of the British School of Archaeology in Jerusalem and the American School of Prehistoric Research under the direction of D. A. E. Garrod (1929–1934), this reconstructed cranium is the smallest of Near Eastern Neandertals and shows typical midfacial projection of this taxon. Although it is archaic-like in some respects, it represents the transition to modern *Homo sapiens*.

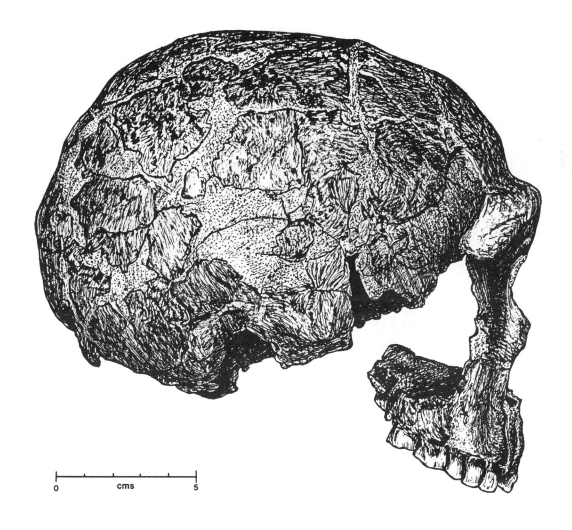

0   cms   5

References: Garrod and Bate, 1937; McCown and Keith, 1939; Higgs and Brothwell, 1961; Brothwell, 1961; Jelinek, 1982; Grün, 1993; Conroy, 1997; Wolpoff, 1998.

Gibraltar 1

Specimen: **Gibraltar 1**

Geographic location: Forbes' Quarry, Gibraltar

Taxonomic affiliation: *Homo sapiens*

Dating: Upper Pleistocene (45,000–70,000 yBP)

General description: This cranium was found during construction of military fortifications in the mid-nineteenth century. Although the specimen was reported to the Gibraltar Scientific Society soon after discovery and donated to the Museum of the Royal College of Surgeons (London) twenty years after discovery, it received no major attention by the scientific world until the turn of the century. The right side of the cranial vault and most of the face are present and show characteristic Neandertal features. Compared with earlier hominids, this individual has a reduced face, but with a large increase in the size of the nasal aperture. Note the large, irregularly rounded eye orbits, low forehead, and protruding occipital.

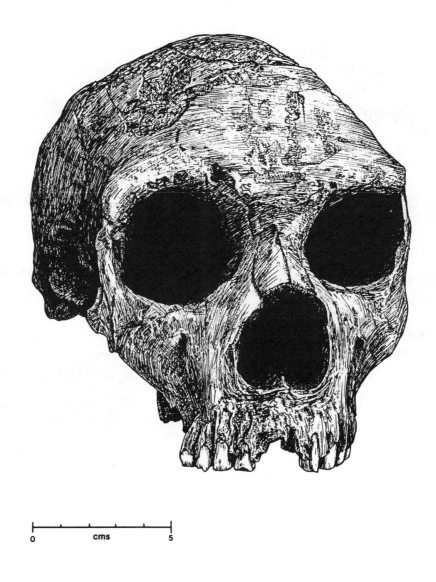

0   cms   5

References: Busk, 1865; Keith, 1911; Hrdlička, 1930; Brace and Montagu, 1977; Reader, 1981; Skinner and Sperber, 1982; Stringer, Hublin, and Vandermeersch, 1984.

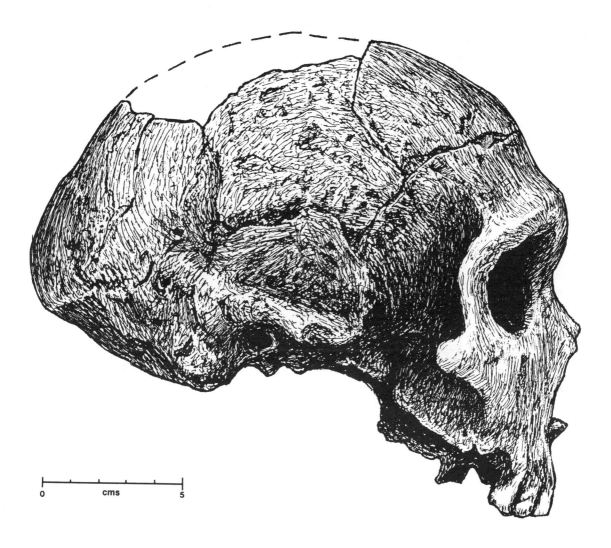

0   cms   5

Krapina 3

Specimen: **Krapina 3 (Cranium C)**
Geographic location: Krapina, Croatia

Taxonomic affiliation: *Homo sapiens*
Dating: Upper Pleistocene (130,000 yBP)

General description: Probably the best known cranium from Krapina—the most complete from the site—this is one of a series of hundreds of hominid bones and teeth recovered by Dragutin Gorjanovic-Kramberger (Croatian National Museum and University of Zagreb) from excavations of a rock shelter during the 1899–1905 field seasons. This adult female cranium is comprised of most of the upper face; the base, rear portion, and the lower face are absent. The face is light in construction—note the small zygomas—and broad in overall size. The morphology of the browridges and the wide distance between eye orbits are typically Neandertal. The presence of cutmarks on this and other individuals from the site suggests that they may have been defleshed with stone tools prior to their deposition in the shelter. Study of dental growth arrest markers known as hypoplasias indicates that the Krapina Neandertal population experienced elevated levels of physiological stress perhaps caused by periodic food shortages. These stress levels, however, are not remarkable in comparison with those of recent humans.

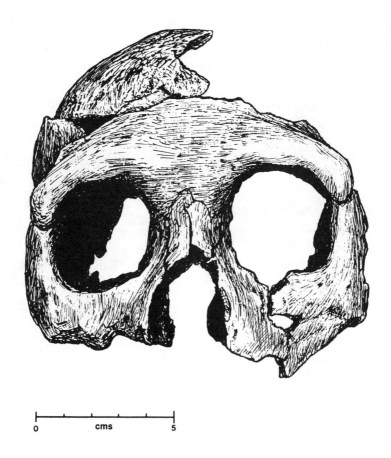

0   cms   5

References: Gorjanović-Kramberger, 1906; Hrdlička, 1930; Smith, 1976a, 1976b; 1980, 1984; Trinkaus, 1975; Wolpoff, 1979, 1980; Ullrich, 1986; Allsworth-Jones, 1986; White, 1986; Smith and Smith, 1986; Frayer and Russell, 1987; Radovčić, 1985; Radovčić, Smith, Trinkaus, and Wolpoff, 1988; Molnar and Molnar 1985; Ogilvie, Curran, and Trinkaus, 1989; Rink, Schwarcz, Smith, and Radovčić, 1995; Hutchinson, Larsen, and Choi, 1997.

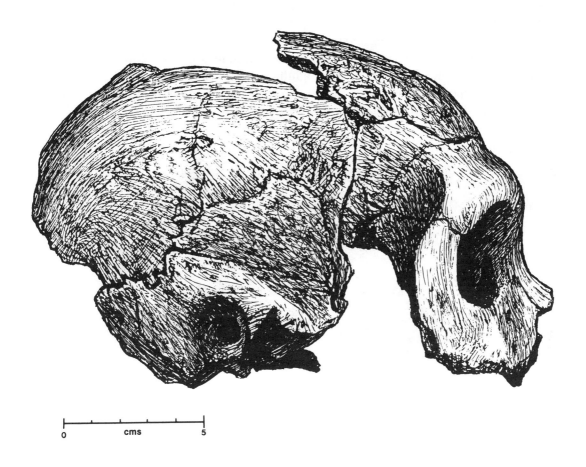

0   cms   5

La Chapelle-aux-Saints

Specimen: **La Chapelle-aux-Saints**
Geographic location: La Chapelle-aux-Saints, France

Taxonomic affiliation: *Homo sapiens*
Dating: Upper Pleistocene (40,000 yBP)

General description: This skull and associated postcranial skeleton were recovered by A. and J. Bouyssonie and L. Bardon from a cave in 1908. The subsequent description of the La Chapelle-aux-Saints human skeletal remains by Marcellin Boule (1911–13) had a tremendous impact on the placement of Neandertals in human evolution. Because of the great influence that Boule's description had on paleoanthropology, Neandertals were seen as brutish, stupid, and having walked with an awkward, slouched gait. Neandertals were perceived as a dead end in the evolution of humankind. Later study of the La Chapelle skeleton by Straus and Cave (1957) and Trinkaus (1985) has demonstrated that the conclusions drawn by Boule do not match the anatomical facts.

The skull exhibits Neandertal traits of forward projection of the midfacial region, occipital bunning, very large nasal aperture and eye orbits, and reduction in size of the zygomas and lower face in comparison with those of earlier hominids. In this specimen, reduction in size of the lower face may be in part related to the fact that this individual was an older adult who in life had experienced extensive tooth loss. As a result, both the upper and lower jaws remodeled such that this region of the face is smaller than what would be expected for an individual who had experienced no premortem tooth loss. The endocranial capacity is large (1600 cc).

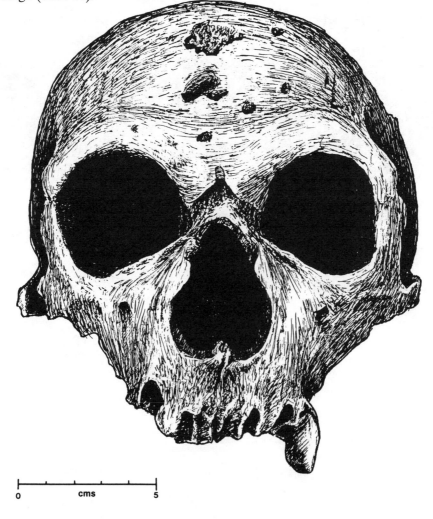

References: Boule, 1911-13; Hrdlička, 1930; Straus and Cave, 1957; Brace, 1964; Kennedy, 1975; Brace and Montagu, 1977; Trinkaus, 1982b, 1985; Stringer, Hublin, and Vandermeersch, 1984; Tappen, 1985; Lewin, 1987; Wolpoff, 1998.

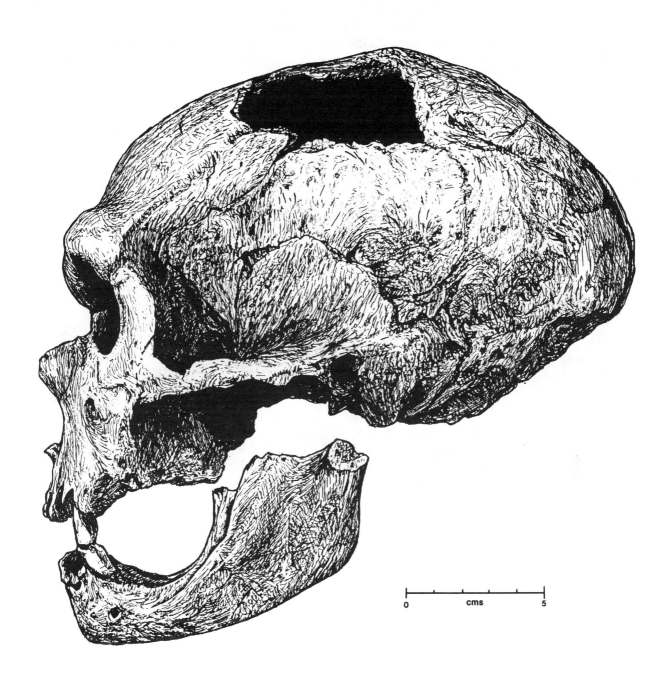

0   cms   5

La Ferrassie 1

Specimen: **La Ferrassie 1**
Geographic location: Savignac du Bugue,
    Dordogne, France

Taxonomic affiliation: *Homo sapiens*
Dating: Upper Pleistocene (40,000–50,000 yBP)

General description: This individual, represented by a nearly complete skeleton that was recovered in 1909 by D. Peyrony and L. Capitan, shows very typical western European Neandertal characteristics—large nasal aperture and large eye orbits. Relative to earlier *Homo sapiens*, the forehead is more filled out and the facial bones are somewhat reduced in size. The cranium, in general, shows more elongation than is typically seen in *Homo sapiens*.

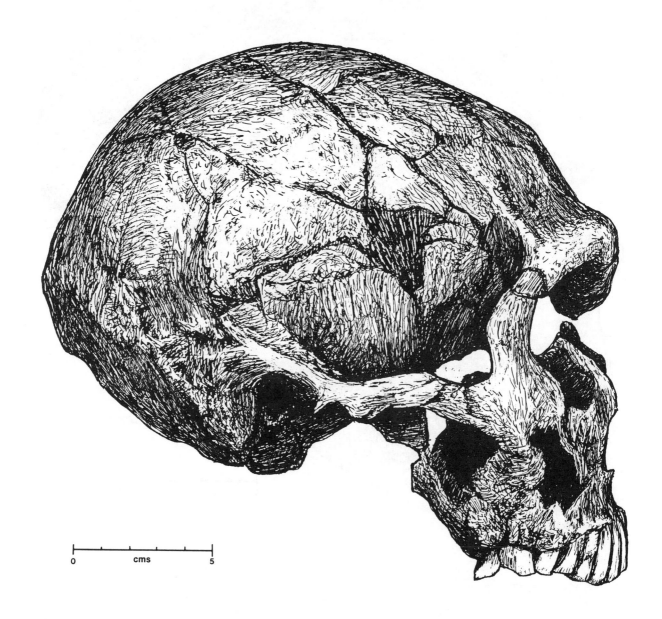

0   cms   5

References: Capitan and Peyrony, 1909; Hrdlička, 1930; Helm, 1976, 1982; Wolpoff, 1998.

La Quina 5

Specimen: **La Quina 5**
Geographic location: Gardes-le-Pontaroux,
 Charente, France

Taxonomic affiliation: *Homo sapiens*
Dating: Upper Pleistocene (40,000–55,000 yBP)

General description: Discovered in 1911 by Henri Martin, this individual is represented by a cranial vault with part of the face and mandible as well as a number of postcranial elements. The cranium shows a series of Neandertal traits: large, irregularly rounded eye orbits, small zygomas, broad nasal aperture, large anterior dentition, and a pronounced occipital bun.

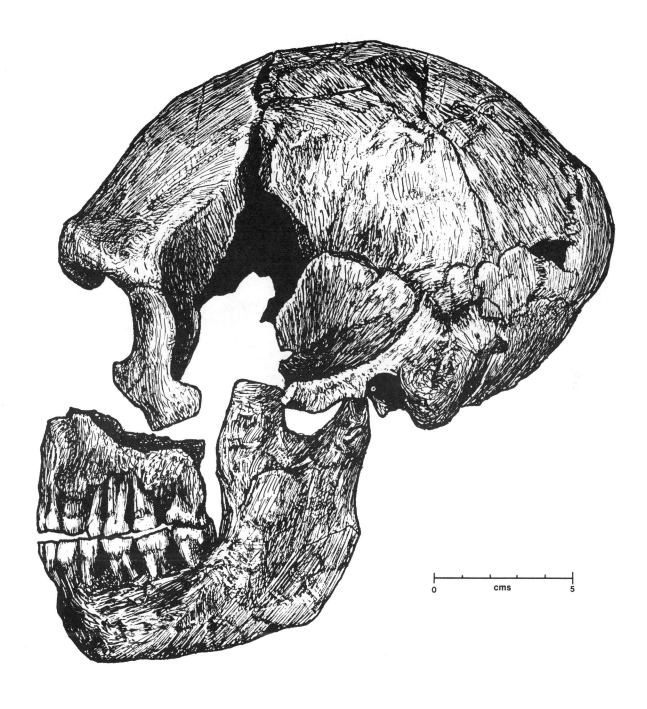

0   cms   5

References: Martin, 1923; Hrdlička, 1930; Wolpoff, 1998.

147

La Quina 18

Specimen: **La Quina 18**
Geographic location: Gardes-le-Pontaroux, Charente, France

Taxonomic affiliation: *Homo sapiens*
Dating: Upper Pleistocene (40,000–55,000 yBP)

General description: Discovered in 1915 by H. Martin, this complete cranium of a seven-year-old child is an especially important fossil in that it shows developmental features of a Neandertal prior to adulthood. In particular, the cranium tends to look modern—it exhibits very small browridges, a globular vault, and a relatively small face. However, those features that are associated with the robustness typical of Neandertals have not yet developed in this individual. Note, though, the very wide nose and large (partially erupted) central incisors that are characteristic of Neandertals.

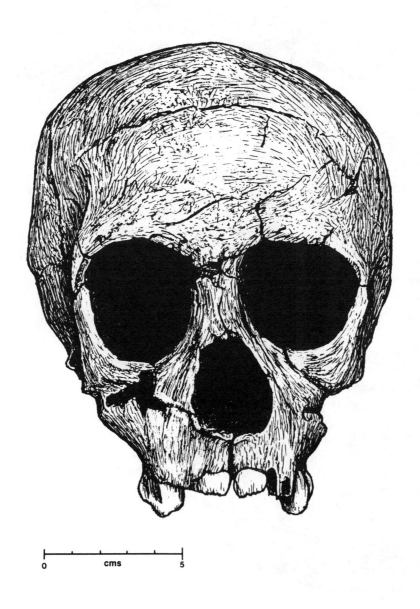

References: Martin, 1923; Skinner and Sperber, 1982; Wolpoff, 1998.

Mt. Circeo 1

Specimen: **Mt. Circeo (Guattari) 1**
Geographic location: San Felice Circeo,
  Latina, Italy

Taxonomic affiliation: *Homo sapiens*
Dating: Upper Pleistocene (40,000–60,000 yBP)

General description: This mostly complete cranium was recovered in the Guattari Cave in 1939. The morphology is typically Neandertal, especially in the projection of the mid-face and development of the zygomas and lower face in general. The face is quite large and has a large nasal aperture. The cranium is more modern appearing than those of the contemporaries to the north (France, Germany, eastern-central Europe), a factor that probably reflects less cold adaptation in the region.

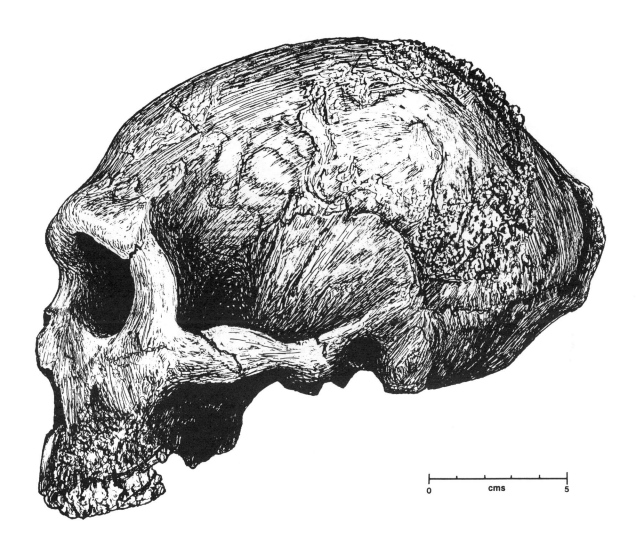

References: Blanc, 1939; Sergi, 1939, 1948, 1958; Stringer, Hublin, and Vandermeersch, 1984; Stiner, 1991, 1994; White and Toth, 1991; Wolpoff, 1998.

Saccopastore 1

Specimen: **Saccopastore 1**
Geographic location: Rome, Italy

Taxonomic affiliation: *Homo sapiens*
Dating: Upper Pleistocene (60,000 yBP)

General description: This specimen was recovered in 1929 by S. Sergi. Like many Neandertals, the cranium is very long and low, and has a large nasal aperture and large eye orbits. Unlike other specimens of this hominid group, however, the back portion of the cranium is rounded. Although the browridges are broken away, if present they would likely have been quite small. The Saccopastore cranium shares with the Mt. Circeo cranium more modern features that probably reflect a reduced expression of traits that have been associated with cold adaptation in western and eastern-central Europe.

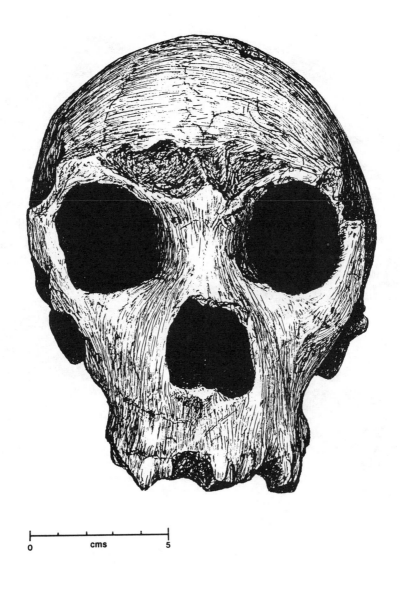

0   cms   5

References: Sergi, 1931, 1948, 1958; Stringer, Hublin, and Vandermeersch, 1984; Condemi 1985; Wolpoff, 1998.

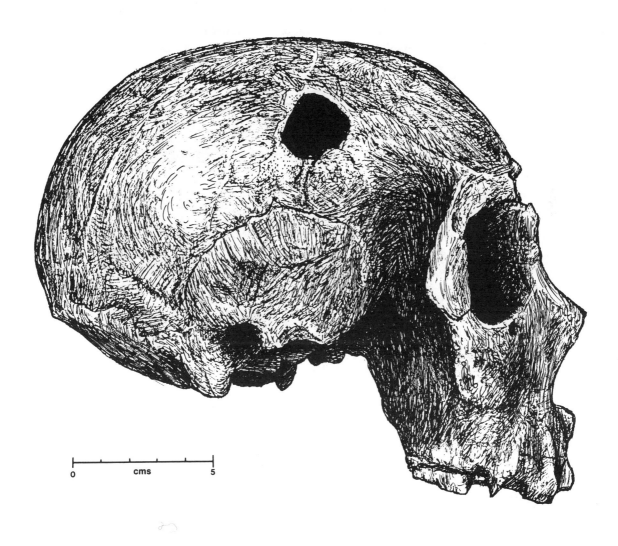

0 cms 5

Neandertal

Specimen: **Neandertal**
Geographic location: Hochdahl, Germany

Taxonomic affiliation: *Homo sapiens*
Dating: Upper Pleistocene (35,000–70,000 yBP)

General description: This specimen is one of the first fossil hominids to receive attention by the scientific community. The cranium and partial skeleton were recovered in 1856 by limestone quarrymen in the process of clearing a cave. Along with the greater part of an associated skeleton, the cranium passed into the hands of a local teacher who in turn showed it to the anatomist H. Schaaffhausen. An initial description of the fossil was presented in a scientific meeting in 1857 by Schaaffhausen. Within a short time, the eminent nineteenth-century evolutionary biologist, Thomas Henry Huxley, examined and described the cranium. Huxley recognized the similarity of the cranium to crania from a number of recent human populations, and in light of the large endocranial capacity, he did not believe that this individual represented an early ancestor of humankind, but rather, an extreme end in human variation from much later in antiquity. However, the overall morphology of the cranium—especially the contour of the browridges, top and back—is consistent with the western European Neandertal pattern. DNA extracted from a bone sample of the humerus reveals key differences between this hominid and modern *Homo sapiens*.

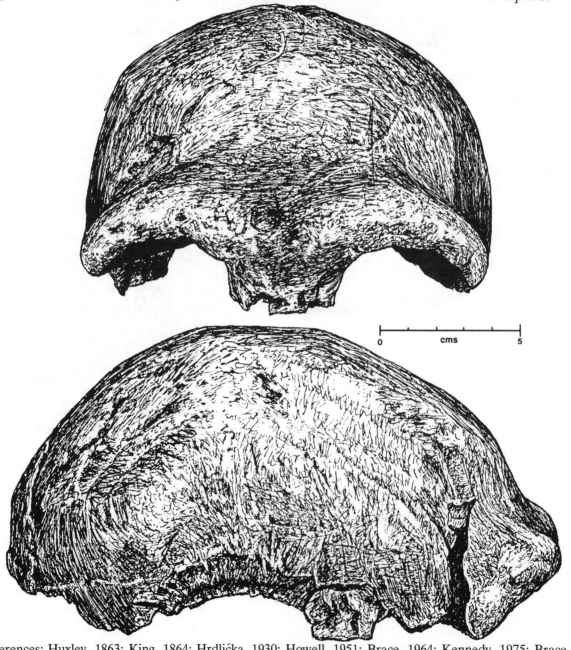

References: Huxley, 1863; King, 1864; Hrdlička, 1930; Howell, 1951; Brace, 1964; Kennedy, 1975; Brace and Montagu, 1977; Reader, 1981; Smith, 1984; Krings, Stone, Schmitz, Krainitzki, Stoneking, and Pääbo, 1997; Ward and Stringer, 1997; Wolpoff, 1998.

Modern *Homo sapiens*

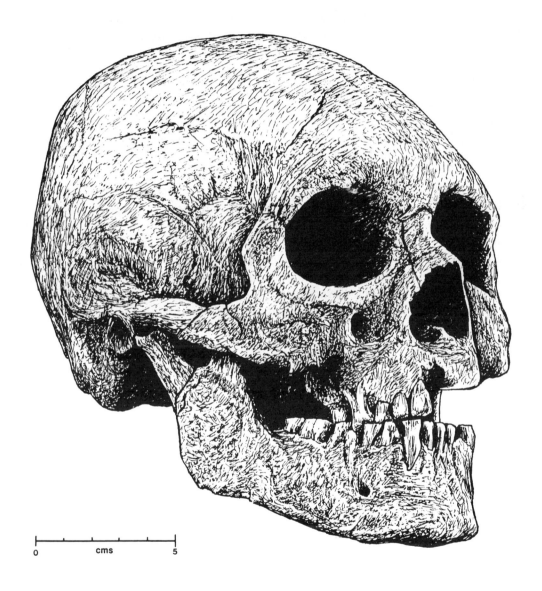

SDM 16704

8. MODERN *HOMO SAPIENS*

Late Pleistocene-Holocene Hominids

Modern *Homo sapiens* make their worldwide appearance just over 30,000 years ago. In some areas of the globe—most notably in Africa—anatomically modern humans appear by 60,000 and perhaps 100,000 years ago (see Smith, Falsetti, and Donnelly, 1989). The comparison of earlier late Pleistocene modern *Homo sapiens* with later archaic *Homo sapiens* reveals a number of important trends. In particular, cranial anatomy is more gracilized, with smaller faces and jaws and teeth. The cranium has a more filled out, rounded contour. This period of time shows very rapid cultural and biological changes. The tempo of change in this regard appears to accelerate at an ever increasing rate.

The geographical expansion and population size increase during this period is unprecedented. It is during this time that we see the spread of humanity into previously uninhabited continents, including Australia, North America, and South America. Australia was first inhabited by modern *Homo sapiens* some 50,000 to 60,000 years ago (Roberts, Jones, and Smith 1990). The expansion of human population into the Americas occurred via the Bering land bridge from northeastern Asia, an event which occurred between 11,000 and 13,000 years before the present (Owen, 1984; Larsen, 1985a; Grayson, 1988; Lynch, 1990; Dillehay and Meltzer, 1991; Meltzer, 1995; Larsen and Patterson, 1997).

The last 10,000 years has been a period of dramatic and rapid dietary changes in the evolution of the Hominidae. It is during this period of time that we see the shift in subsistence economy from that based entirely on wild plants and animals to that based on domesticated plants and animals. This change had far-reaching implications for human settlement and consequent alteration in health and disease patterns (Larsen, 1995). The shift in dietary emphasis has also resulted in alterations in food-preparation technology and food consistency. The decreasing demands on the faces and jaws in this regard ultimately resulted in further reductions in the size of faces, jaws, and teeth and an overall reorientation of the craniofacial anatomy (Carlson and Van Gerven, 1979; Calcagno, 1989). In a similar vein, the postcranial skeleton shows a remarkable reduction in robusticity in modern humans compared with those of earlier hominids, which is almost certainly due to declining workloads and activity generally (Ruff and others, 1993).

In comparison with late archaic *Homo sapiens*, modern humans have a smaller brain size (average=1450 cc). However, accounting for body size changes—Neandertals outweighed modern humans by about 30%—modern humans have somewhat larger brains (Ruff and others, 1997; Gibbons, 1997).

The hominids remains that are represented in this time period include the following:

1. Wadi el-Mughara, Israel: Skhul 5
2. Nazareth, Israel: Jebel Qafzeh 6
3. Lothagam, Tanzania: Lo. 4b
4. Aswan, Egypt: Wadi Kubbaniya
5. Wadi Halfa, Sudan: Wadi Halfa 25
6. Kerma, Sudan: Kerma 27
7. Oued Agrioun, Algeria: Afalou 13
8. Dordogne, France: Combe Capelle
9. Dordogne, France: Cro-Magnon 1
10. Premosti, Czech Republic: Predmost 3
11. Dunstable, England: Five Knolls 18
12. Gushikami, Okinawa: Minatogawa I
13. New South Wales, Australia: Tandou
14. Victoria, Australia: Kow Swamp 1
15. Tulungaguna, Indonesia: Wadjak 1
16. Browns Valley, Minnesota, U.S.A.: Browns Valley
17. Del Mar, California, U.S.A.: SDM 16704
18. Tepexpan, Mexico
19. Patagonia, Chile: Cerro Sota 2
20. Nevada, U.S.A.: Humboldt Sink
21. St. Catherines Island, Georgia, U.S.A.: SCDG K102/76

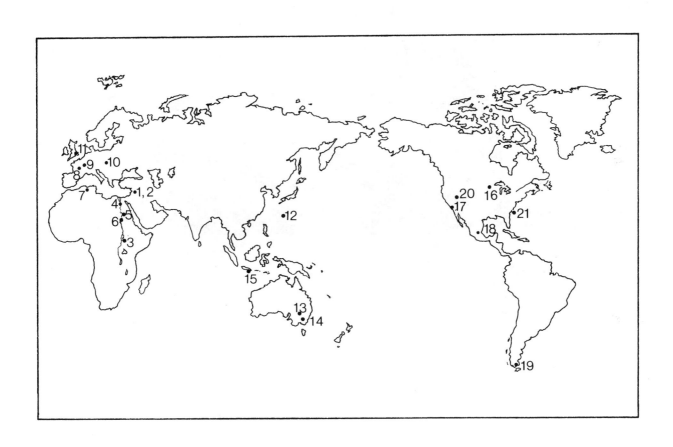

Skhul 5

Specimen: **Skhul 5**
Geographic location: Wadi el-Mughara, Israel

Taxonomic affiliation: *Homo sapiens*
Dating: Upper Pleistocene (31,000–33,000 yBP; 80,000–120,000 yBP)

General description: Recovered by T. D. McCown and H. Movius, Jr., in 1932, and subsequently restored by C. E. Snow, this nearly complete cranium with mandible and dentition shows modern features. The teeth are small, but the dimensions are similar to those of Neandertals. The forward projecting mid-facial region is also Neandertal-like. However, a hallmark of modern *Homo sapiens*—the chin—is an obvious feature of this individual, and there is a lack of bunning of the occipital. If the Middle Pleistocene dates are correct, then this modern hominid may represent evidence for co-existence of early modern humans with Neandertal populations (e.g., Tabun) in the same region.

0   cms   5

References: Garrod and Bate, 1937; Keith and McCown, 1937; McCown and Keith, 1939; Snow, 1953; Higgs and Brothwell, 1961; Brothwell, 1961; Stringer, Grün, Schwarcz, and Goldberg, 1989; Marshall, 1990; Mercier, Valladas, Bar-Yosef, Vandermeersch, Stringer, and Joron, 1993; Grün, 1993; Wolpoff, 1998.

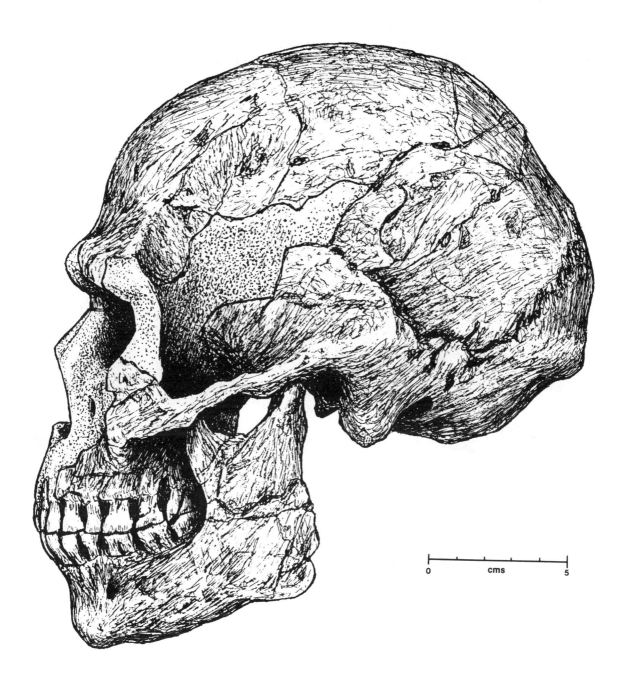

0   cms   5

Jebel Qafzeh 6

Specimen: **Jebel Qafzeh 6**
Geographic location: Nazareth, Israel

Taxonomic affiliation: *Homo sapiens*
Dating: Upper Pleistocene (30,000–55,000 yBP; 92,000–115,000 yBP)

General description: This individual—one of at least 15 from the site—was recovered in the mid-1930s by R. Neuville and M. Stekelis from Qafzeh. It consists of a nearly complete cranium with the dentition. This cranium is archaic with respect to its large eye orbits, nasal aperture, and anterior dentition. The high forehead and generally rounded vault is similar to Skhul 5, thus giving it a very modern appearance. Given these modern traits, a number of workers have suggested that this individual and others from the site are, in fact, fully modern *Homo sapiens*. If this is a modern *Homo sapiens* and the Early Upper Pleistocene dates are correct, then an argument for contemporaneous occupation of this region of the world by both late archaic and modern human populations is more tenable.

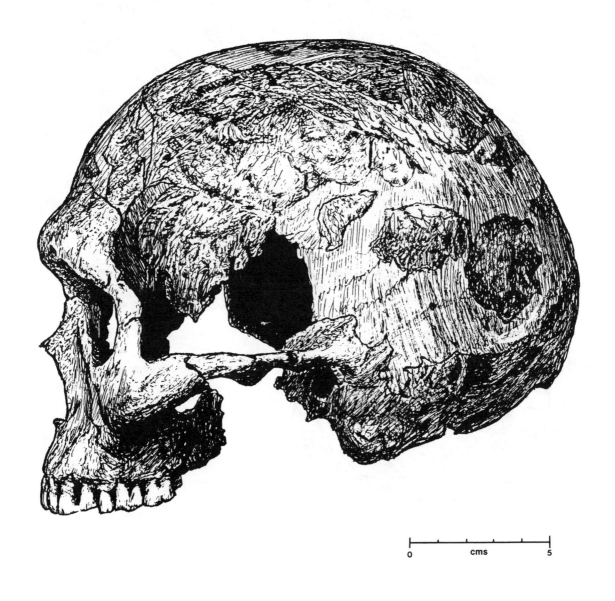

0 cms 5

References: Koppel, 1935; Howell, 1958; Vandermeersch, 1972, 1981; Vallois and Vandermeersch, 1972; Mann and Trinkaus, 1973; Bar Yosef and Vandermeersch, 1981, 1993; Trinkaus, 1983; Allsworth-Jones, 1986; Stringer, 1988; H. Valladas, Reyss, Joron, G. Valladas, Bar-Yosef, and Vandermeersch, 1988; Schwarcz, Grün, Vandermeersch, Bar-Yosef, Valladas, and Tchernov, 1988; Mellars, 1989; Smith, Falsetti, and Donnelly, 1989; Marshall, 1990; Rak, 1990; Grün, 1993; Wolpoff, 1998.

Lo. 4b

Specimen: **Lo. 4b**
Geographic location: Lothagam,
  West Turkana, Kenya

Taxonomic affiliation: *Homo sapiens*
Dating: Early Holocene (6000–9000 yBP)

General description: This individual is one of nearly 30 individuals that were recovered as surface and *in situ* finds during intermittent field seasons from 1965 to 1975 by L. H. Robbins, B. M. Lynch, and others from Michigan State University. The skeletal materials were subsequently restored and studied by T. W. Phenice and J. L. Angel. This skull is the best preserved from the Lothagam series. It is narrow, long, and large, but not particularly robust. The posterior cranium is rounded, and the face is wide with distinct alveolar prognathism. The mandible is large. This specimen and associated remains from the site are similar to, but with relatively greater robusticity than, a number of Subsaharan populations presently occupying the region.

0   cms   5

References: Angel, Phenice, Robbins, and Lynch, 1980; Schepartz, 1987; Wolpoff, 1998.

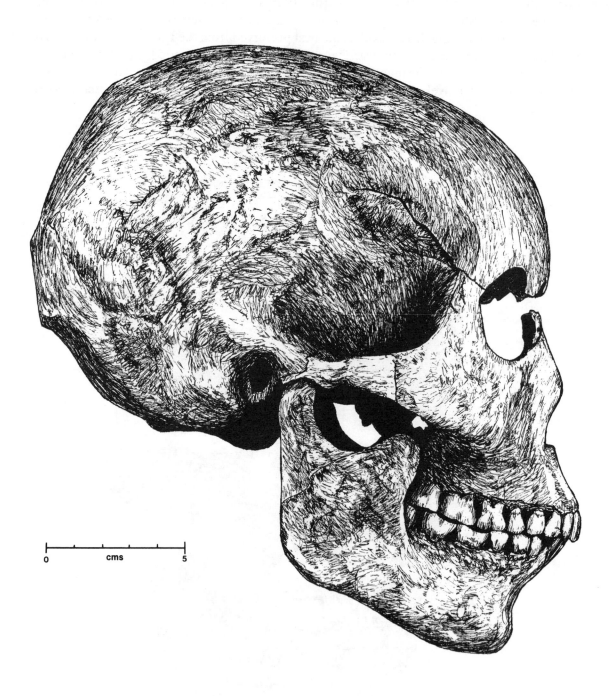

0 cms 5

Wadi Kubbaniya

Specimen: **Wadi Kubbaniya**
Geographic location: Wadi Kubbaniya, Aswan, Egypt

Taxonomic affiliation: *Homo sapiens*
Dating: Terminal Pleistocene–early Holocene (8,000–20,000 yBP)

General description: This specimen, found in association with a partial postcranial skeleton, was discovered by a Southern Methodist University expedition under the direction of Fred Wendorf in 1982. The cranium is missing most of the posterior vault. It was prepared by T. D. Stewart and M. Tiffany and subsequently described by J. L. Angel and J. O. Kelley at the National Museum of Natural History, Smithsonian Institution. This young, adult male has a robust face with some flare of the zygomas, more so than seen in living humans in the region today. The lower facial prognathism is marked, and the chin is not especially prominent. The surfaces of the lumbar vertebrae show concavities reflecting a very strenuous, physically demanding lifeway. Unusual development of the shoulder, elbow, forearm, and hand indicate that this individual was probably quite adept at throwing and thrusting spears. He had recovered from two injuries, including trauma to the right forearm and elbow. The latter injury resulted from a spear wound.

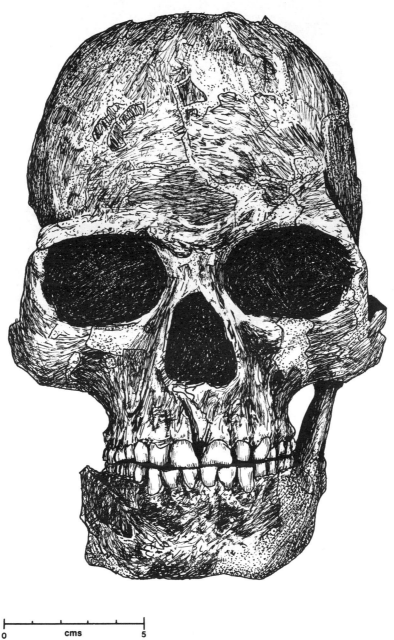

References: Stewart, 1985; Stewart and Tiffany, 1986; Angel and Kelley, 1986; Wendorf, Schild, Close, and others, 1988.

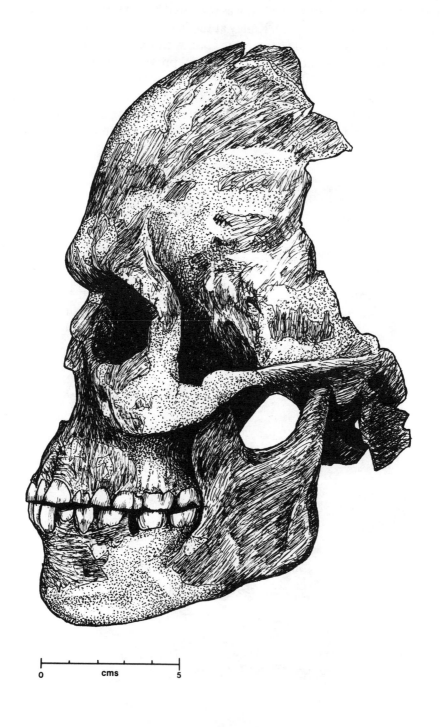

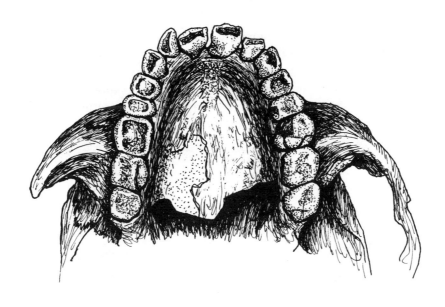

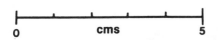

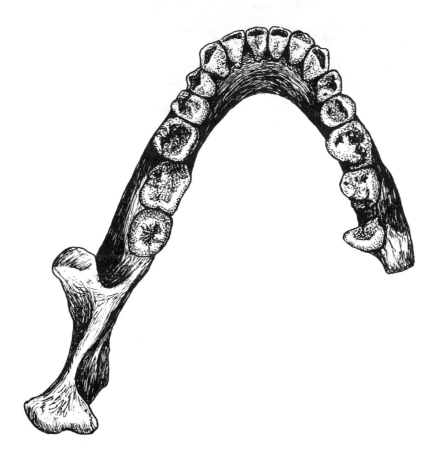

Wadi Halfa 25

Specimen: **Wadi Halfa 25**

Geographic location: Wadi Halfa, Sudan (Nubia)

Taxonomic affiliation: *Homo sapiens*

Dating: Terminal Pleistocene–early Holocene (6,000–12,000 yBP)

General description: Wadi Halfa 25 is one of nearly 40 individuals excavated during the 1963–1964 University of Colorado Nubian Expedition by G. Armelagos, E. Ewing, and D. Greene. The site (6B36) is important in that it contained the remains of individuals who practiced a hunting and gathering lifeway just prior to the development of an agricultural based economy in Nubia. Relative to later populations in the region, the cranial series shows well developed masticatory apparatus, large supraorbitals, alveolar prognathism, and large, morphologically complex teeth.

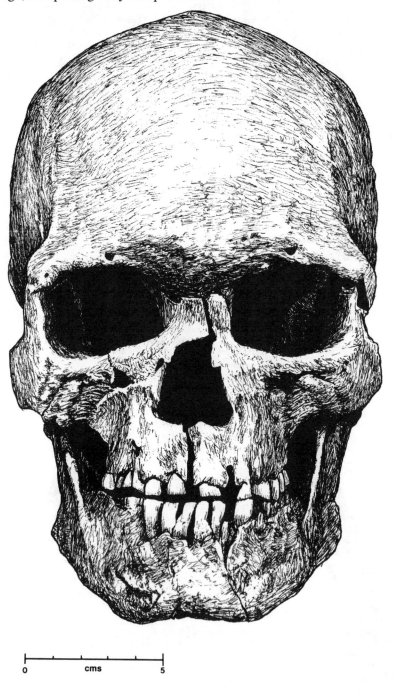

References: Hewes, Irwin, Papworth, and Saxe, 1964; Greene, Ewing, and Armelagos, 1967; Armelagos, 1969; Saxe, 1971; Greene and Armelagos, 1972.

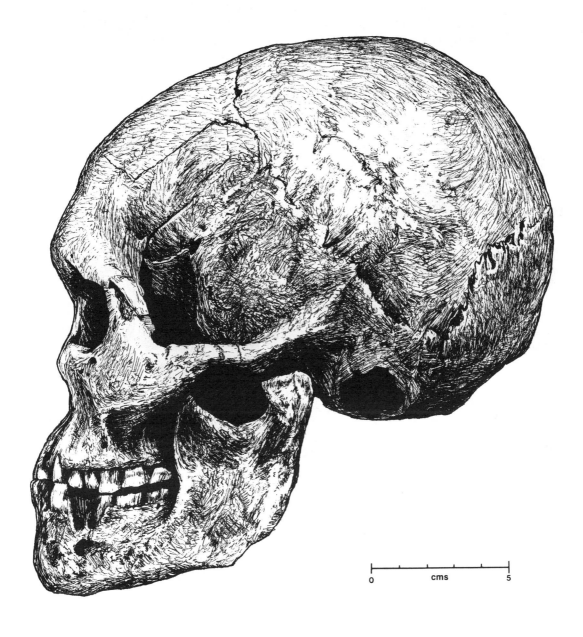

Kerma 27

Specimen: **Kerma 27**
Geographic location: Kerma, Sudan (Nubia)

Taxonomic affiliation: *Homo sapiens*
Dating: Recent Holocene (3500 yBP)

General description: This skull is one of several hundred recovered from a cemetery along the Nile River by G. A. Reisner (Harvard University-Boston Museum of Fine Arts joint expedition) during the 1913–1916 field seasons. Comparison of earlier Nubian crania shows that the crania from the Kerma series are more rounded and shorter, the areas of muscle attachment are smaller, and overall, there is a reorientation of the face and cranial vault. D. S. Carlson and others have suggested that these changes arose consequent to the adoption of softer foodstuffs associated with an agricultural-based diet.

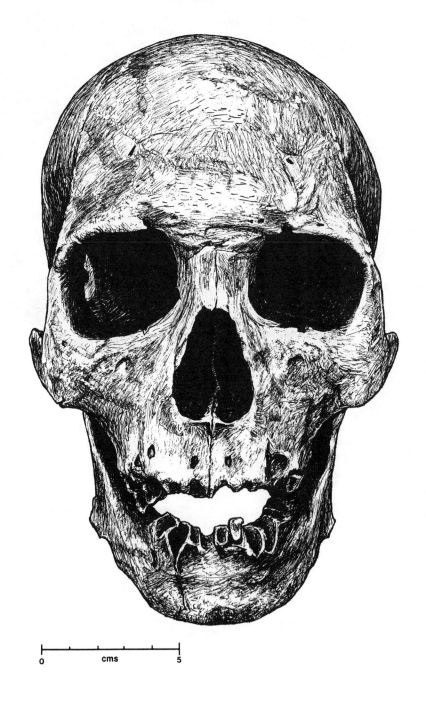

0   cms   5

References: Reisner, 1923; Collett, 1933; Trigger, 1976; Adams, 1977; Carlson, 1976; Carlson and Van Gerven, 1977, 1979; Martin, Armelagos, Goodman, and Van Gerven, 1984.

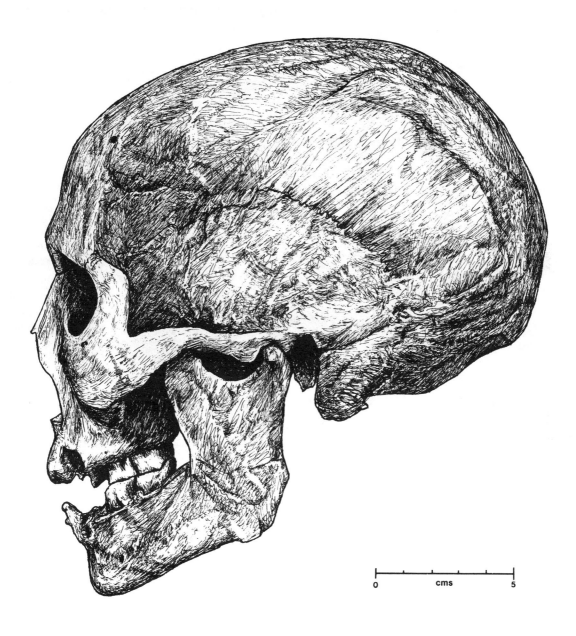

0   cms   5

Afalou 13

Specimen: **Afalou 13**
Geographic location: Oued Agrioun, Algeria

Taxonomic affiliation: *Homo sapiens*
Dating: Terminal Pleistocene–early Holocene (8,000–12,000 yBP)

General description: This individual is one of at least 50 individuals recovered by C. Arambourg in the late 1920s, thus representing an important sample of modern humans from north Africa. Although robust, the skull is that of a modern human that is ancestral to populations occupying northern Africa today. Note the shortened face, small mandible and zygoma, and more rounded facial skeleton relative to those of earlier hominids.

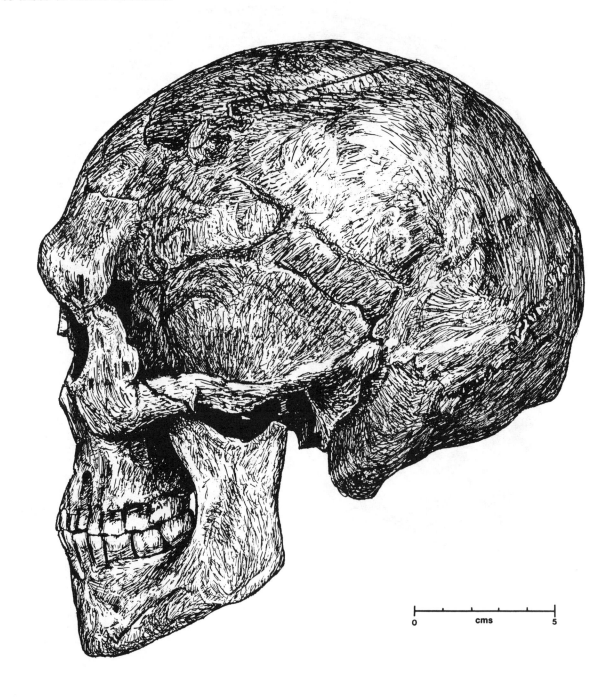

References: Arambourg, 1929; Arambourg, Boule, Vallois, and Verneau, 1934; Anderson, 1968; Greene and Armelagos, 1972; Wolpoff, 1998.

Combe Capelle

Specimen: **Combe Capelle**
Geographic location: Saint Avit Senieur,
  Dordogne, France

Taxonomic affiliation: *Homo sapiens*
Dating: Upper Pleistocene (30,000–35,000 yBP)

General description: This skull, along with most of an associated postcranial skeleton, was recovered by O. Hauser and his workmen in 1909. Although dating has been somewhat problematical—indeed the skeleton may be an from an intrusive burial dating much later in time—this individual is from the earliest levels of the Upper Paleolithic in western Europe. The cranium and most of the rest of the skeleton did not survive the Allied bombing of Berlin at the end of World War II. Relative to Neandertals, the cranium is higher and narrower, but the face still shows some forward projection. The mandible is short and the chin is not pronounced.

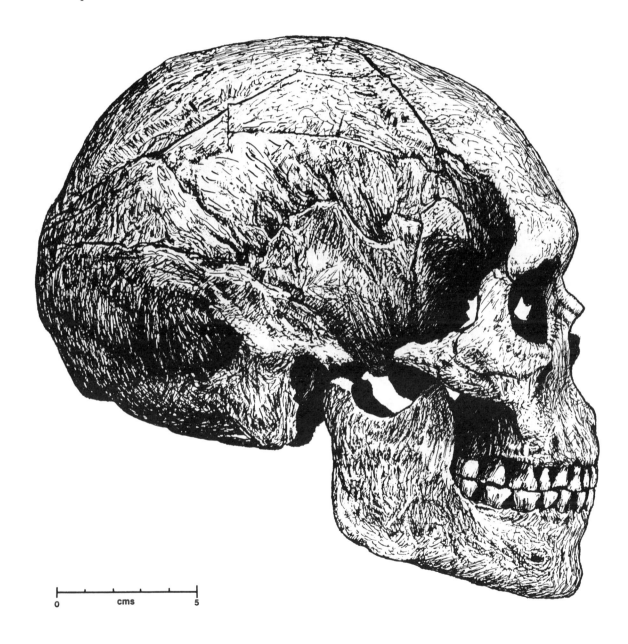

References: Klaatsch and Hauser, 1909; Von Bonin, 1935; Frayer, 1978, Stringer, Hublin, and Vandermeersch, 1984; Gambier, 1989; Dibble and Lenoir, 1995; Wolpoff, 1998.

Cro-Magnon 1

Specimen: **Cro-Magnon 1**

Geographic location: Les Eyzies, Dordogne, France

Taxonomic affiliation: *Homo sapiens*

Dating: Upper Pleistocene (23,000–27,000 yBP)

General description: Popularly known as the "Old Man" of Cro-Magnon, this individual was discovered in 1868 by workmen and removed from the site by L. Lartet along with the remains of three other adults and four children. This skull exhibits a series of features that characterize the emergence of fully modern *Homo sapiens* in western Europe. Note in particular the broad, high face, the high, vertical forehead, small browridges, relatively narrow nasal aperture, and prominent chin. The projecting nasal bones are reminiscent of the earlier archaic *Homo sapiens* of western Europe (compare with La Chapelle-aux-Saints). The lower face is quite small. In part, the diminutive size of the lower face may be associated with reduction in facial bone mass that is often associated with the aging process, particularly in older adults.

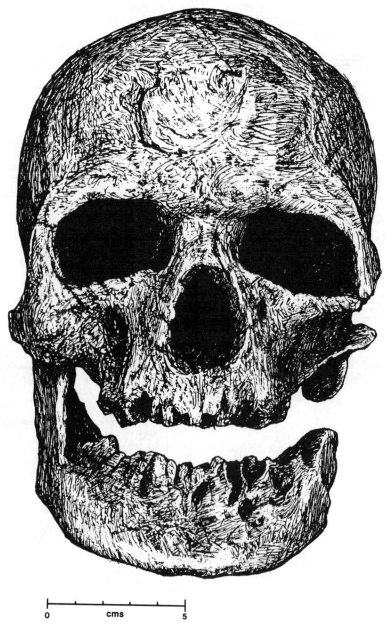

References: Vallois and Billy, 1965; Camps and Olivier, 1970; Brace and Montagu, 1977; Frayer, 1978; Stringer, Hublin, and Vandermeersch, 1984; Gambier, 1989; Wolpoff, 1998.

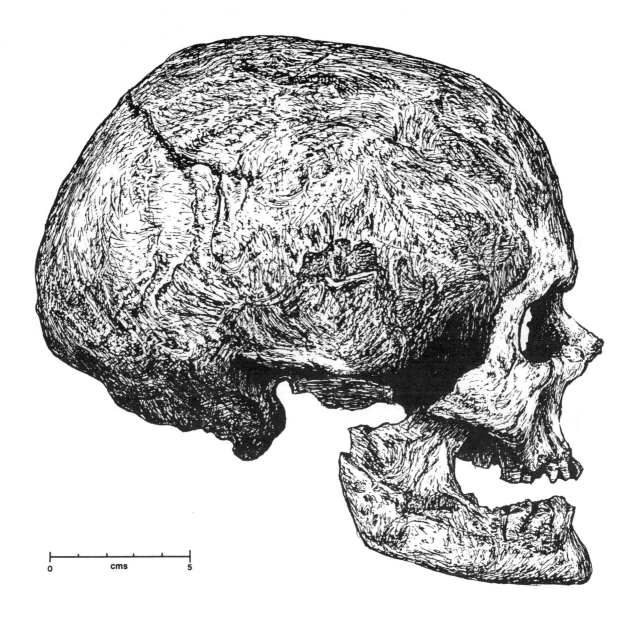

0   cms   5

Predmost 3

Specimen: **Predmost 3**
Geographic location: Predmosti, Moravia, Czech Republic

Taxonomic affiliation: *Homo sapiens*
Dating: Upper Pleistocene (26,000 yBP)

General description: Excavated by K. J. Maska in 1894, this specimen along with the remains of at least 26 other skeletal individuals was destroyed during the closing days of World War II. The specimen illustrated below shows features that characterize fully modern *Homo sapiens*—high, fully domed cranium, reduced midfacial projection and face in general, and a well-developed chin. Although modern, this individual possesses archaic-like features (compare with Skhul 5), such as morphology of the browridges, occipital bunning, and fairly well-developed cranial base spongy bone. The comparison of dental size with earlier hominids shows marked reduction.

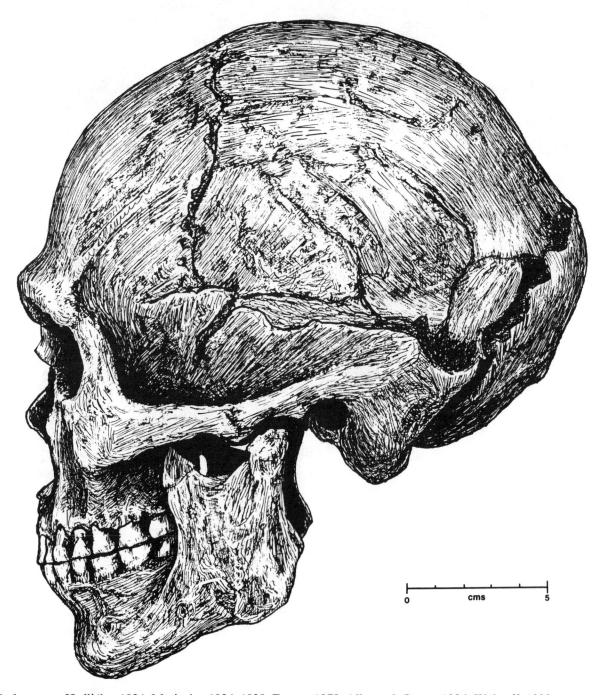

References: Hrdlička, 1924; Matiegka, 1934, 1938; Frayer, 1978; Allsworth-Jones, 1986; Wolpoff, 1998.

Five Knolls 18

Specimen: **Five Knolls 18**
Geographic location: Dunstable, England

Taxonomic affiliation: *Homo sapiens*
Dating: Recent Holocene (1500–3500 yBP)

General description: This skull is from a series of human remains that were excavated by the University College London during the 1925–1929 field seasons. The site is a barrow with a central tomb and associated secondary and tertiary burials. The cranium shows features that are characteristic of fully modern Europeans. In comparison with Cro-Magnon, the Five Knolls cranium is similar in overall morphology, but the zygomas are considerably reduced and the cranial vault is higher and more rounded.

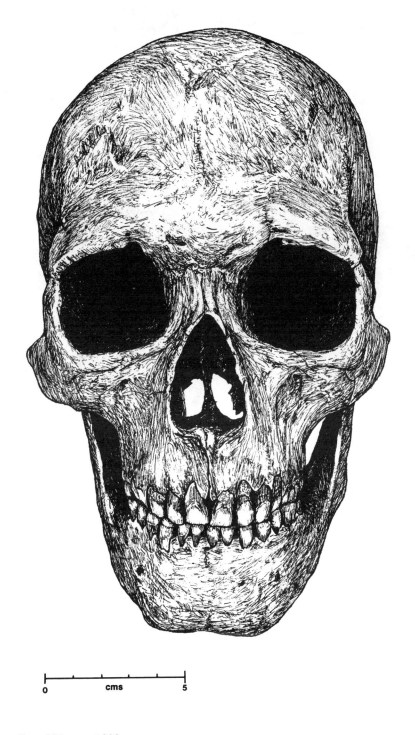

0   cms   5

References: Dingwall and Young, 1933.

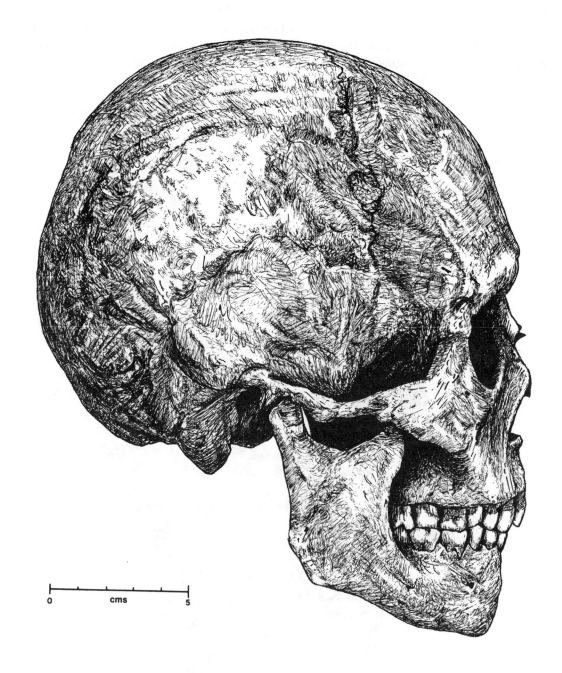

0   cms   5

Minatogawa 1

Specimen: **Minatogawa 1**
Geographic location: Gushikami, Okinawa

Taxonomic affiliation: *Homo sapiens*
Dating: Late Pleistocene (18,250 yBP)

General description: This is one skull from the series of five to nine individuals that were recovered by an interdisciplinary scientific team from Japan during the years 1968 to 1974. The nearly complete cranium exhibits thick cranial bones with relatively large browridges, low, broad face, wide nose, and well-developed zygomas. Tooth size of this individual is similar to that of *Homo sapiens* of the Upper Paleolithic and considerably larger than that of modern Japanese. Given the paucity of upper Pleistocene hominid remains in Asia, this skull and other materials from Minatogawa help to provide a better understanding of human evolution from this area of the globe.

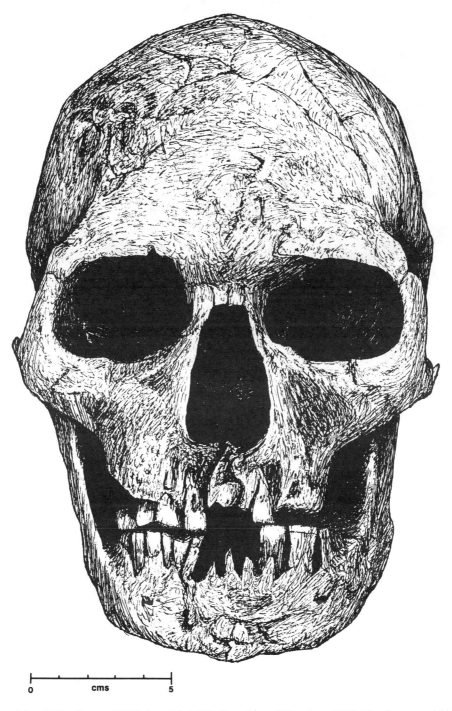

0   cms   5

References: Suzuki and Hanihara, 1982; Suzuki, 1982; Suzuki and Tanabe, 1982; Hanihara and Ueda, 1982.

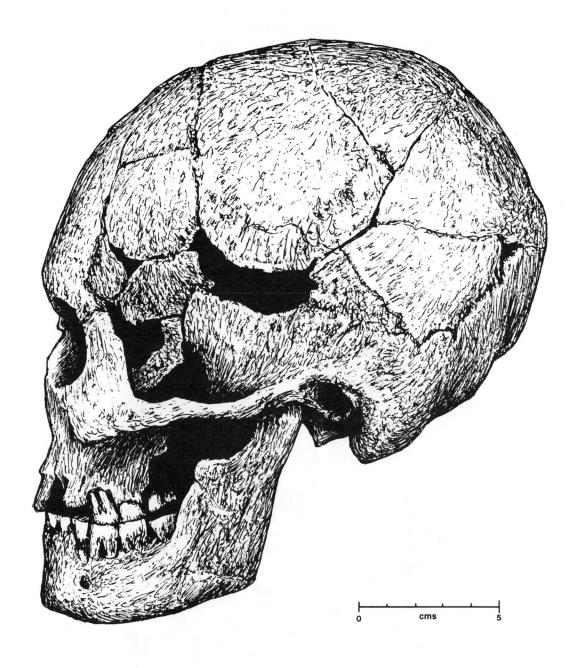

0   cms   5

Tandou

Specimen: **Tandou**

Geographic location: Bootingee Station, New South Wales, Australia

Taxonomic affiliation: *Homo sapiens*

Dating: Late Pleistocene (15,000 yBP)

General description: This specimen was discovered in 1967 by Duncan Merrilees (Western Australian Museum). It is a partial cranium with most of the left side of the face missing; parts of the right side of the face are damaged or have been lost. Part of the dentition is present. The browridge is weakly developed, but the lower face projects and, like the skulls of modern Australian native populations, the area of the eye orbits is quite broad. The forehead is filled out, and it is nearly vertical. Although the specimen is badly eroded and damaged, it is an important representative of early Aborigines from this continent. Its general robusticity and form are features exhibited in recent Australian Aborigines.

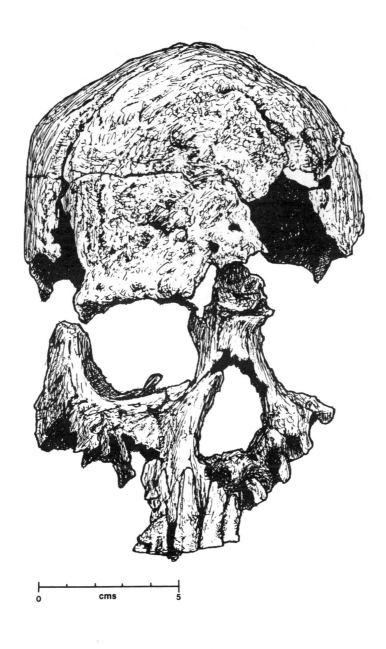

0   cms   5

References: Merrilees, 1973; Freedman and Lofgren, 1983.

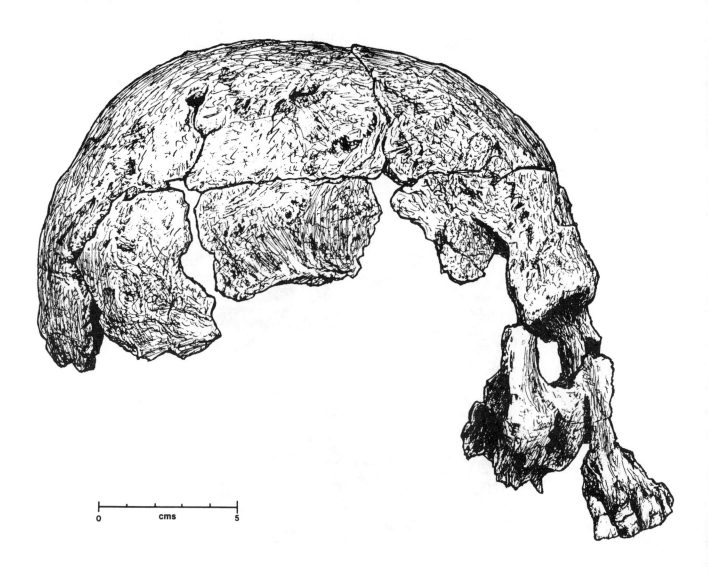

0   cms   5

Kow Swamp 1

Specimen: **Kow Swamp 1**
Geographic location: Leitchville,
 Victoria, Australia

Taxonomic affiliation: *Homo sapiens*
Dating: Terminal Pleistocene–early Holocene
 (9,000–13,000 yBP)

General description: This cranium was recovered as part of a series of more than 22 individuals in the central Murray River Valley by A. G. Thorne and A. L. West in 1968. This specimen is one of two relatively complete crania from the site. It displays archaic features in that it is large and robust, the face projects, the forehead is quite low, and the supraorbitals are well developed. A number of similarities between the Kow Swamp cranial series and *Homo erectus* in Asia—in particular, Indonesian *Homo erectus*—suggests a continuity of human population in Australasia.

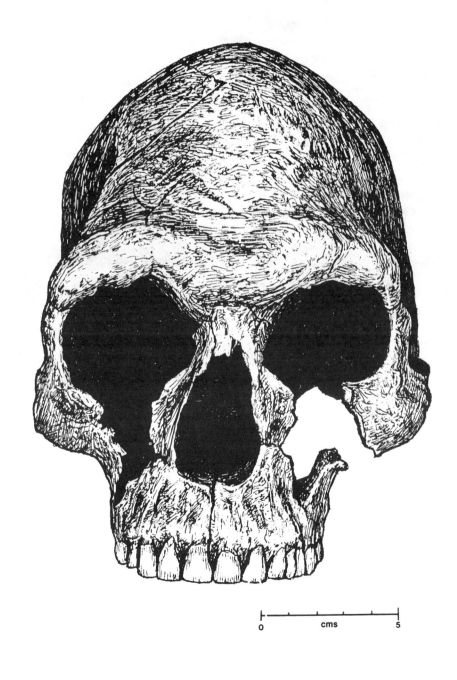

References: Thorne, 1971, 1972, 1975; Thorne and Macumber, 1972; Thorne and Wolpoff, 1981; Wolpoff, Wu, and Thorne, 1984; Freedman and Lofgren, 1983; Brown, 1987; Wolpoff, 1998.

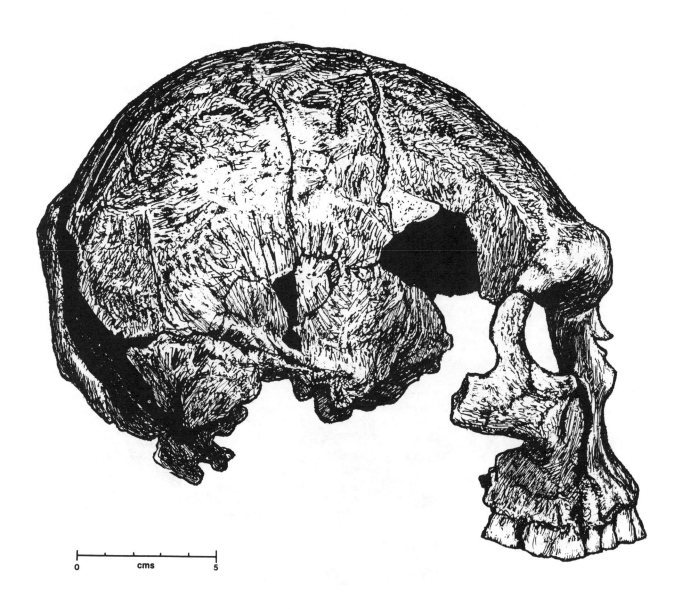

0 cms 5

Wadjak 1

Specimen: **Wadjak 1**
Geographic location: Tulungagung, Indonesia

Taxonomic affiliation: *Homo sapiens*
Dating: Terminal Pleistocene–early Holocene
(10,000–2,000 yBP)

General description: This nearly complete cranium was found by B. D. van Reitschoten in 1889. The browridge shows pronounced development, but only in the area above the nose. The face in general is large and displays very obvious forward projection, especially beneath the nose. The cranium has features that are seen in both Australian native populations and late Pleistocene Chinese.

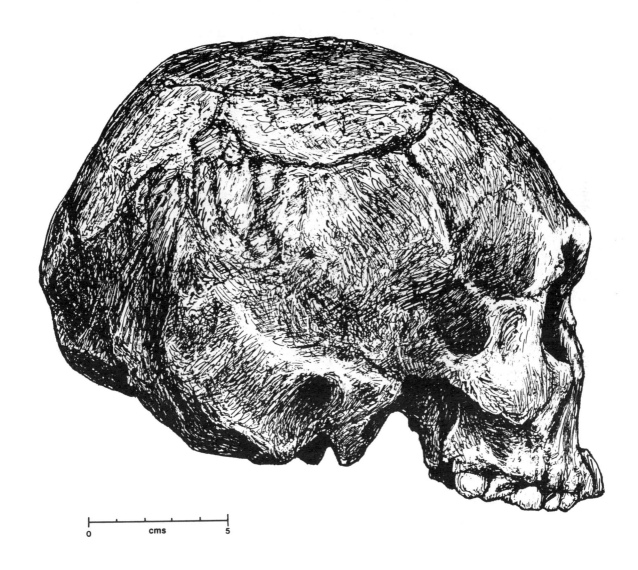

0   cms   5

References: Dubois, 1922; Pinkley, 1935–6; von Koenigswald, 1956; Wolpoff, Wu, and Thorne, 1984; Wolpoff, 1998.

Browns Valley

Specimen: **Browns Valley**
Geographic location: Browns Valley,
 Traverse County, Minnesota, U.S.A.

Taxonomic affiliation: *Homo sapiens*
Dating: Early Holocene (8,700 yBP)

General description: This cranium and an associated fragmentary skeleton was discovered in 1933 by William H. Jensen and M.L. Granoski in a burial pit intruding into Pleistocene gravels. The association with Paleoindian lithics is consistent with a relatively early date. The skull was reconstructed by Albert E. Jenks. Although some facial elements are missing, the overall completeness allows an informed perspective on some of the earliest well-documented Native Americans, the Paleoindians. The cranium is long and, overall, quite robust. The supraorbital torus is well developed, and the face and jaws are markedly broad with well defined attachments areas for the masticatory musculature. In comparison with recent American Indians, Paleoindians have longer and narrower cranial vaults and shorter and narrower faces.

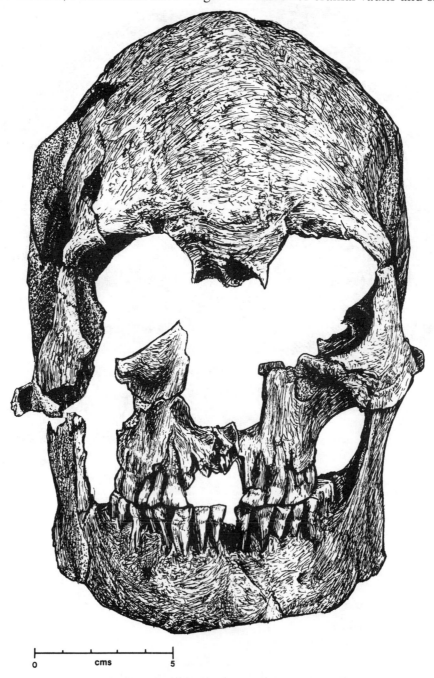

0   cms   5

References: Jenks, 1937; Hrdlička, 1937; Protsch, 1978; Steele and Powell, 1992, 1993; Jantz and Owsley, 1997.

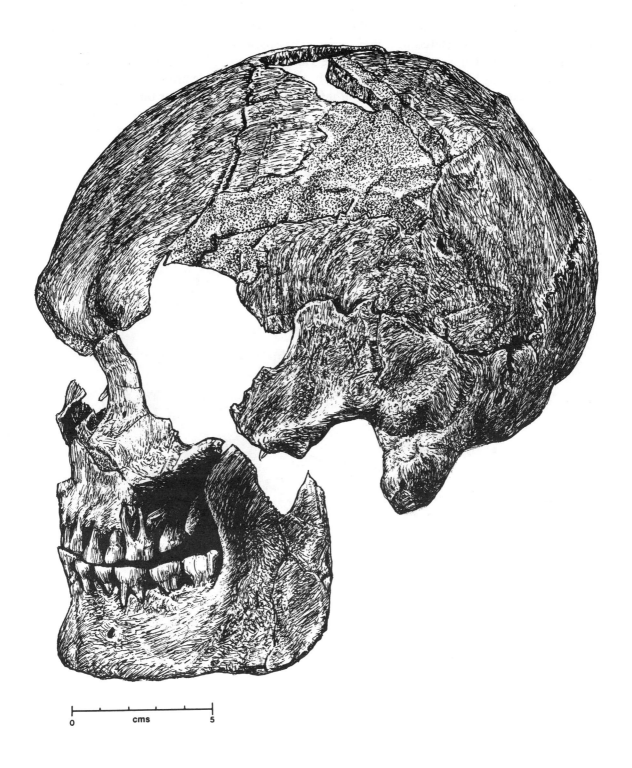

SDM 16704

Specimen: **SDM 16704**
Geographic location: Del Mar, California, U.S.A.

Taxonomic affiliation: *Homo sapiens*
Dating: Terminal Pleistocene–middle Holocene (11,800 yBP; 4900 yBP)

General description: This skull is one of a group of human remains recovered in California that have been suggested to be associated with Pleistocene deposits. Redating of this specimen has shown, however, that the individual is most likely from the Holocene epoch. The remains of this individual were recovered in 1929 by M. J. Rogers (San Diego Museum of Man) while eroding out of a midden of undetermined cultural context. The cranium is of a fully modern *Homo sapiens* but exhibits a low forehead and a well-developed face. These features, however, are within the range of variation in recent Native Americans. S. L. Rogers has noted that the morphology is similar to that of later aboriginal crania recovered from the southern California and Baja California coasts.

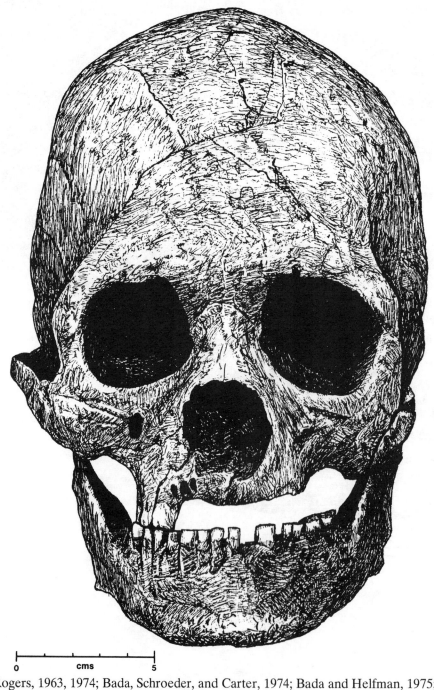

0 cms 5

References: Rogers, 1963, 1974; Bada, Schroeder, and Carter, 1974; Bada and Helfman, 1975; Taylor and Payen, 1979; Smith, 1977; Bischoff and Rosenbauer, 1981; Owen, 1984; Bada, Gillespie, Gowlett, and Hedges, 1984; Taylor, Payen, Prior, and others, 1985; Marshall, 1990.

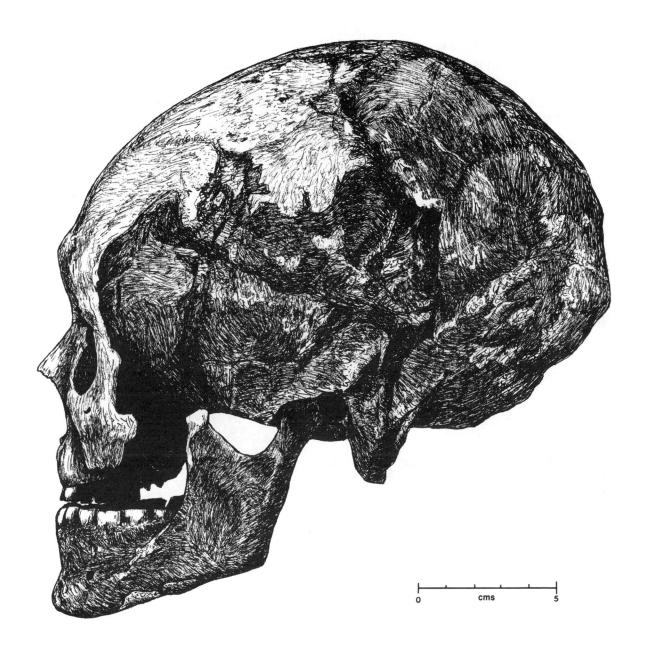

0   cms   5

Tepexpan

Specimen: **Tepexpan**
Geographic location: Tepexpan, Mexico

Taxonomic affiliation: *Homo sapiens*
Dating: Terminal Pleistocene–early Holocene
(11,000 yBP)

General description: Along with the remains of a nearly complete skeleton, this skull was recovered by Hellmut de Terra in 1947. It is one of a very limited number of human remains representative of the first human occupation of the New World. The cranium is robust with large browridges, and the face is short and broad. The back of the cranium is quite rounded, as is the vault in general. The overall configuration of the skull is completely modern and is virtually indistinguishable from that of recent Native Americans. Although in apparent temporal association with mammoth bones, recent analyses suggest the possibility that this individual may represent a later intrusion.

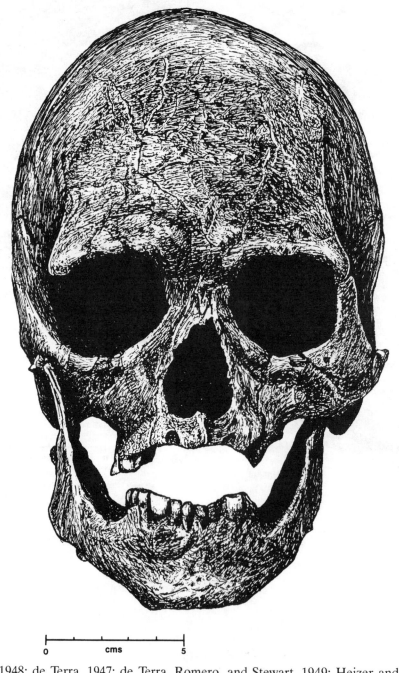

0   cms   5

References: Field, 1948; de Terra, 1947; de Terra, Romero, and Stewart, 1949; Heizer and Cook, 1959; Moss, 1960; Genovés, 1960; Smith, 1976c; Owen, 1984.

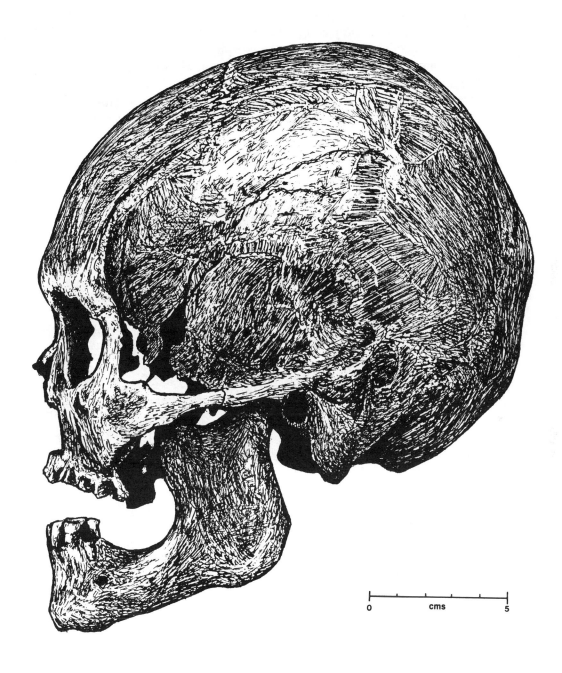

Cerro Sota 2

Specimen: **Cerro Sota 2**
Geographic location: Patagonia, Chile

Taxonomic affiliation: *Homo sapiens*
Dating: Terminal Pleistocene–early Holocene
(11,000 yBP; 3380 yBP)

General description: This complete cranium was recovered from cave deposits along with a small series of other human remains in 1936 by Junius Bird for the American Museum of Natural History. Reconstruction of this cranium by Harry L. Shapiro presents us with an individual with a long cranium and a small face; the occipital is somewhat projecting. The cranial vault is small and shows small browridges. There is very little facial projection. In all respects, the morphology of the individual is fully modern and is consistent with variability seen in South American Native Americans. A study of dental traits of this individual and others from Cerro Sota Cave and nearby Palli Aike Cave revealed great similarity in dental structure with living and recent American Indians as well as with eastern Asians. Although considered to be a Paleoindian by most authorities, a recently determined AMS ^{14}C date of 3380 yBP suggests a much more recent origin for this individual.

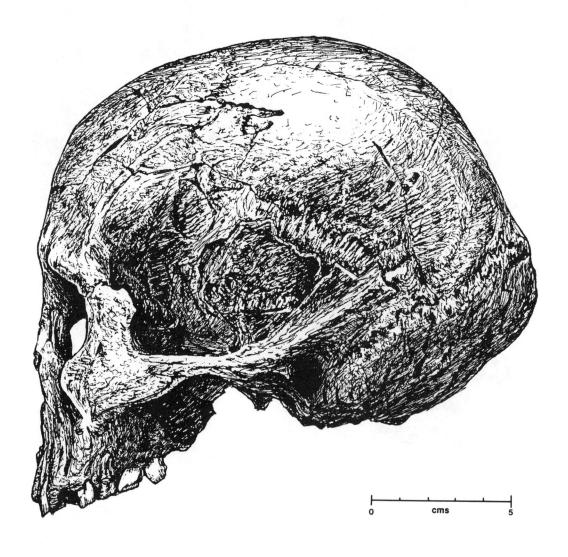

0　　　　　cms　　　　　5

References: Bird, 1938; Patterson and Bird, 1975; Turner and Bird, 1981; Hedges, Housley, Bronk, and Van Klinken, 1992; Massone, 1996.

Humboldt Sink

Specimen: **Humboldt Sink**

Geographic location: Humboldt Sink, Nevada, U.S.A.

Taxonomic affiliation: *Homo sapiens*

Dating: Recent Holocene (<1500 yBP)

General description: This cranium was collected by amateur archaeologists from an unknown prehistoric context. As is generally characteristic of prehistoric western Great Basin hunter-gatherers, this individual is markedly robust. The cranium is long, yet broad and high. The face shows wide, deep and flaring zygomas, features that reflect masticatory muscles of notable size. Other muscle attachment areas on the cranium are also well developed. These features, coupled with extreme dental wear, are suggestive of great demands placed on the chewing complex for both masticatory and extra-masticatory (tool use) functions.

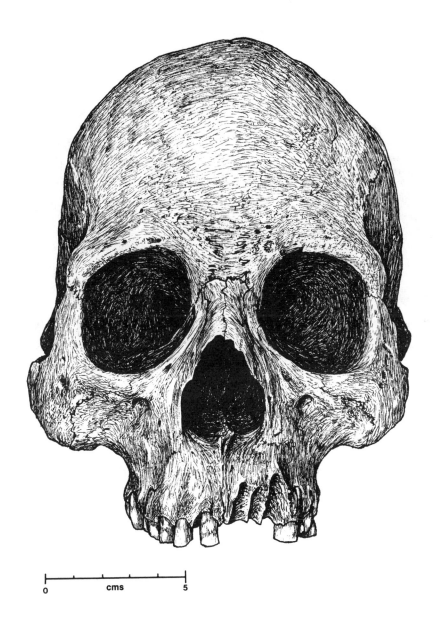

References: Kennedy, 1959; Larsen, 1985b; Larsen and Kelly, 1995.

SCDG K102/76

Specimen: **SCDG K102/76**

Geographic location: St. Catherines Island, Georgia, U.S.A.

Taxonomic Affiliation: *Homo sapiens*

Dating: Recent Holocene (300–400 yBP)

General description: This cranium is from a group of over four hundred individuals excavated from an early historic Native American cemetery—Santa Catalina de Guale—by C. S. Larsen during the 1982–1986 field seasons. Native American populations occupying this area of North America focussed their diet on agriculture (especially maize), and included hunting, gathering, and fishing for the acquisition of supplemental resources. The cranium is gracile with a small face and short, rounded vault. In comparison with the earlier, prehistoric hunting and gathering aboriginal populations on the Georgia coast, the later prehistoric and historic agricultural-focussed populations in the region possess shorter vaults and reduced facial skeleton reflecting consumption of softer foods. This reduction in craniofacial robusticity is part of a worldwide trend.

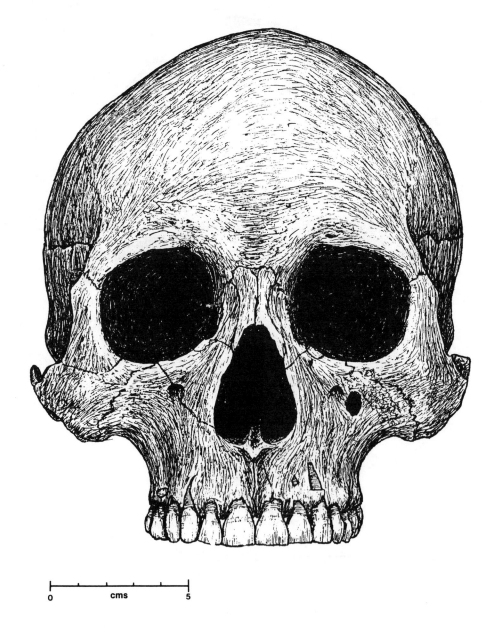

0   cms   5

References: Larsen, 1982, 1984, 1990, 1995; Larsen, Schoeninger, Hutchinson, Russell, and Ruff, 1990; Thomas, 1987.

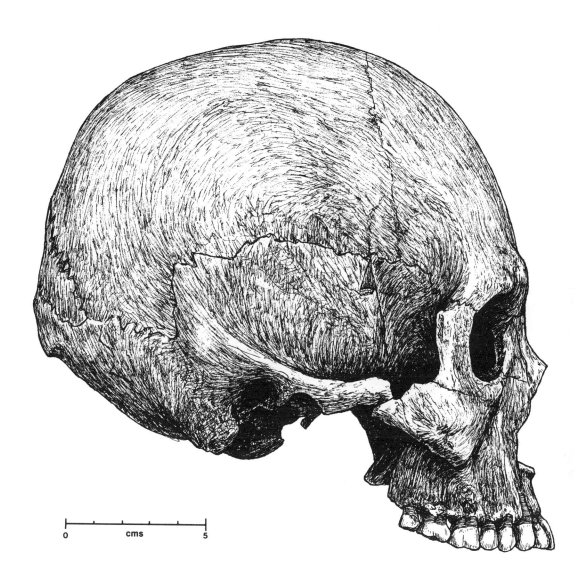

0   cms   5

SOURCE MATERIALS USED FOR FIGURE PREPARATIONS

The table below is a list of source materials, casts, original specimens, and photographs that were used in the preparation of the figures for this volume. With the exception of reference photographs of cranial materials from the Humboldt Sink, Nevada, and Santa Catalina de Guale, Georgia, all photographs are published. Under the headings of "Cast" and "Original," "x" indicates that a cast or original specimen were examined; "–" indicates that neither were used.

Specimen	Cast	Original	Reference Photograph
chimpanzee, male	x	x	–
pygmy chimpanzee, male	x	–	–
gorilla, male	x	x	–
gorilla, female	x	x	–
orangutan, male	x	x	–
chimpanzee, postcrania	–	x	–
modern *Homo*, postcrania	–	x	–
DPC-2803	x	–	–
DPC-1028	x	–	–
KNM-RU 7290	x	–	Le Gros Clark and L. Leakey, 1951; Walker et al. 1983
KNM-WK 16999	–	–	R. Leakey et al., 1988b
UM.P. 62-11	x	–	Simons, 1972
KNM-WK 16950A	–	–	R. Leakey et al., 1988a
M.N.H.N.P.-A.C.36	x	–	–
IGF 11778	–	–	Straus, 1963
AS95-500	–	–	Alpagut et al., 1996
IVPP PA580; IVPP PA 644	–	–	Inst. Vertebrate Pal., 1980; Hammond, 1983
GSP 15000	x	x	Pilbeam, 1982
G. blacki II, III	x	–	–
KNM-MB 20573	–	–	McCrossin and Benefit, 1993
BER I, 1'91	x	–	–
L.H. 4	x	–	White et al., 1981
A.L. 288-1	x	–	Johanson and Edey, 1981; Berge et al., 1984
A.L. 333-63	x	–	Bush et al., 1982
A.L. 200-1a	–	–	White and Johanson, 1982
A.L. 400-1a	–	–	Kimbel et al ., 1982
Hadar composite skull	x	–	Johanson and Edey, 1981; Kimbel et al., 1984; Kimbel and White, 1988
A.L. 444-2	–	–	Johanson and Blake, 1996
KNM-WT 17000	x	–	R. Leakey and Walker, 1988
Taung	x	x	R. Leakey, 1981
STS 5	x	x	Reader, 1981; Rak, 1983
STS 14	x	–	Howell, 1970
STS 52a, STS 52b	x	–	Skinner and Sperber, 1982
STS 71	x	–	Wolpoff, 1998; Rak, 1983
SK 46	x	–	Reader, 1981
SK 23, SK 48	x	–	Wolpoff, 1998; Rak, 1983
O.H. 5	x	–	Tobias, 1967; Rak, 1983
Peninj	x	–	Wolpoff, 1998
KNM-ER 406	x	–	Reader, 1981; R. Leakey and Walker, 1976
KNM-ER 732	x	–	Wolpoff, 1998; Kimbel et al., 1984
KNM-ER 1805	x	–	Walker, 1981

O.H. 13	x	–	Skinner and Sperber, 1982
O.H. 24	x	–	Wolpoff, 1998
O.H. 8	x	–	Napier, 1967
KNM-ER 1813	x	–	R. Leakey and Lewin, 1977
KNM-ER 1470	x	–	R. Leakey, 1973a; R. Leakey and Walker, 1976; R. Leakey and Lewin, 1977
StW 53	x	x	Washburn and Moore, 1980
SK 847	x	–	Clarke and Howell, 1972; Wolpoff, 1998
KNM-ER 3733	x	–	R. Leakey and Lewin, 1977; Wolpoff, 1998
KNM-ER 3883	x	–	Wolpoff, 1998
KNM-WT 15000	x	–	R. Leakey and Walker,1985b; Simons, 1989; Walker and Leakey, 1993; Walker, personal communication
O.H. 9	x	–	Rightmire, 1979
Bodo	–	–	Conroy et al., 1978
Ternifine 2, 3	x	–	Arambourg, 1955a; Wolpoff, 1998
Mauer	x	x	Wolpoff, 1998
Sangiran 4	x	x	Weidenreich, 1940
Sangiran 17	x	–	Thorne and Wolpoff, 1981
Sangiran IX	–	–	Eijgenraam, 1996; Tyler, personal communication
Hexian	–	–	Wu Rukang, 1983; Wolpoff et al., 1984
Zhoukoudian	x	–	Weidenreich, 1943
Broken Hill 1	x	–	Pycraft et al., 1928
Ndutu	x	–	Tattersall and Delson, 1984
Salé	–	x	Jaeger, 1981
Petralona	–	–	Murrill, 1981; Thorne and Wolpoff, 1981
Arago 21	–	x	de Lumley and de Lumley, 1973; de Lumley, 1981
Atapuerca 5	–	–	Johanson and Blake, 1996
Swanscombe	x	–	Wolpoff, 1998
Steinheim	x	x	Weinert, 1936
Narmada	–	–	Sonakia, 1985
Ngandong 7	x	–	Weidenreich, 1951
Dali	–	–	Wu Rukang, 1983
L.H. 18	–	–	Gowlett, 1984
Omo 2	–	–	Day, 1969; Wolpoff, 1998
Jebel Irhoud 1	–	–	Ennouchi, 1962, 1963
Shanidar 1	–	–	Stewart, 1958; Trinkaus, 1983
Amud 1	–	x	Suzuki and Takai, 1970
Tabun C1	–	–	McCown and Keith, 1939; Wolpoff, 1998
Gibraltar 1	x	–	Reader, 1981; Gowlett, 1984
Krapina 3	x	x	Smith, 1976b; Wolpoff, 1998
La Chapelle	x	–	Wolpoff, 1998
La Ferrassie I	x	x	Wolpoff, 1998
La Quina 5	–	x	Weiner, 1971
La Quina 18	–	–	Martin, 1923
Mt. Circeo 1	x	–	Howell, 1970
Saccopastore 1	x	x	Wolpoff, 1998
Neandertal	x	x	Reader, 1981
Skhul 5	x	x	Howell, 1970
Jebel Qafzeh 6	–	–	Vallois and Vandermeersch, 1972
Lo. 4b	–	–	Angel et al., 1980

Wadi Kubbaniya	–	–	Stewart, 1985
Wadi Halfa 25	–	–	Greene and Armelagos, 1972
Kerma 27	–	–	Collett, 1933
Afalou 13	–	–	Weiner, 1971
Combe Capelle	x	–	Wolpoff, 1998
Cro-Magnon 1	x	–	Wolpoff, 1998
Predmost 3	x	–	Lasker and Tyzzer, 1981
Five Knolls 18	–	–	Dingwall and Young, 1933
Minatogawa 1	–	–	Suzuki and Hanihara, 1982
Tandou	–	–	Freedman and Lofgren, 1983
Kow Swamp I	–	–	Thorne and Wolpoff, 1981
Wadjak 1	x	–	Weiner, 1971
Browns Valley	–	–	Jenks, 1937
SDM 16704	–	–	Rogers, 1974; Lasker and Tyzzer, 1981
Tepexpan 1	–	–	de Terra et al., 1949
Cerro Sota 2	–	x	Patterson and Bird, 1975
Humboldt Sink	–	–	Larsen, unpublished
SCDG K102/76	–	x	Larsen, unpublished

References Cited

Ackerman, Sandra
1989 European Prehistory Gets Even Older. Science 246:28–30.

Adam, Karl Dietrich
1985 The Chronological and Systematic Position of the Steinheim Cranium. In (Eric Delson, editor) Ancestors: The Hard Evidence, pp. 272–276. New York: Alan R. Liss.

Adams, William Y.
1977 Nubia: Corridor to Africa. Princeton: Princeton University Press.

Aiello, Leslie, and Christopher Dean
1990 An Introduction to Human Evolutionary Anatomy. San Diego: Academic Press.

Allbrook, David, and W. W. Bishop
1963 New Fossil Hominoid Material from Uganda. Nature 197:1187–1190.

Allsworth-Jones, P.
1986 The Szeletian and the Transition from Middle to Upper Palaeolithic in Central Europe. Oxford: Clarendon Press.

Alpagut, B., Peter Andrews, Mikael Fortelius, John Kappelman, Ilhan Temizsoy, Hurkan Celebi, and William Lindsay
1996 A New Specimen of *Ankarapithecus meteai* from the Sinap Formation of Central Anatolia. Nature 382:349–351.

Anderson, James E.
1968 Late Paleolithic Skeletal Remains from Nubia. In (F. Wendorf, editor) The Prehistory of Nubia, vol. 2, pp. 996–1040. Dallas: Southern Methodist University Press.

Andrews, Peter J.
1971 *Ramapithecus wickeri* mandible from Fort Ternan, Kenya. Nature 231:192–194.
1982 Hominoid Evolution. Nature 295:185–186.
1983 The Natural History of *Sivapithecus*. In (R. L. Ciochon and R. Corruccini, editors) New Interpretations of Ape and Human Ancestry, pp. 441–464. New York: Plenum Press.
1985 Family Group Systematics and Evolution among Catarrhine Primates. In (E. Delson, editor) Ancestors: The Hard Evidence, pp. 14–22. New York: Alan R. Liss.
1992a Evolution and Environment in the Hominoidea. Nature 360:641–646.
1992b An Ape from the South. Nature 356:106.

Andrews, Peter, and J. E. Cronin
1982 The Relationships of *Sivapithecus* and *Ramapithecus* and the Evolution of the Orang-utan. Nature 297:541–546.

Andrews, Peter, and Yolanda Fernandez Jalvo
1997 Surface Modifications of the Sima de los Huesos Fossil Humans. Journal of Human Evolution 33:191–217.

Andrews, Peter J., and Lawrence Martin
1987 Cladistic Relationships of Extant and Fossil Hominoids. Journal of Human Evolution 16:101–118.

Andrews, Peter, and I. Tekkaya
1980 A Revision of the Turkish Miocene Hominoid *Sivapithecus meteai*. Paleontology 23:85–95.

Andrews, Peter, and Alan Walker
1976 The Primate and Other Fauna from Fort Ternan, Kenya. In (Glynn Isaac and Elizabeth McCown, editors) Human Origins—Louis Leakey and the East African Evidence, pp. 279–304. Menlo Park: W. A. Benjamin.

Angel, J. Lawrence, and Jennifer Olsen Kelley
1986 Description and Comparison of the Skeleton. In (Fred Wendorf and Romuald Schild) The Wadi Kabbaniya Skeleton: A Late Paleolithic Burial from Southern Egypt, pp. 53–70. Dallas: Southern Methodist University Press.

Angel, J. L., T. W. Phenice, L. H. Robbins, and B. M. Lynch
1980 Lopoy and Lothagam No. 2: Late Stone Age Fishermen of Lothagam, Kenya. Michigan State University Anthropological Series 3.

Ankel-Simons, Friderun
1983 A Survey of Living Primates and Their Anatomy. New York: Macmillan Publishing Co.

Arambourg, C.

1929   Découverte d'un Ossuaire Humain du Paléolithique Supérieur en Afrique du Nord. L'Anthropologie 39:219–221.

1955a   A Recent Discovery in Human Paleontology: *Atlanthropus* of Ternifine (Algeria). American Journal of Physical Anthropology 13: 191–201.

1955b   Deuxième Supplément a la Notice sur les Travaux Scientifiques. II. Paléontologie Humaine, pp. 11–13. Paris: Sennac.

Arambourg, C., M. Boule, H. Vallois, and R. Verneau

1934   Les Grottes paléolithiques des Beni-Segoual (Algérie). Archives de l'Institut de Paléontologie Humaine, Mémoire 13.

Arambourg, C., and Y. Coppens

1968   Découverte d'un Australopithécien Nouveau dans les Gisements de l'Omo (Ethiopia). South African Journal of Science 64:58–59.

Arambourg, C., and R. Hoffstetter

1954   Découverte en Afrique du Nord de Restes Humains du Paléolithique Inférieur. Comptes Rendus des Séances de l'Academie des Sciences, Paris 239:72–74.

Armelagos, George J.

1969   Disease in Ancient Nubia. Science 163:255–259.

Arsuaga, Juan Luis, José María Bermúdez, and Eudald Carbonell

1997   Preface (Sima de los Huesos, Atapuerca, Spain). Journal of Human Evolution 33:105–108.

Arsuaga, Juan Luis, Ignacio Martínez, Ana Gracia, José-Miguel Carretero, and Eudald Carbonell

1993   Three New Human Skulls from the Sima de los Huesos Middle Pleistocene site in Sierra de Atapuerca, Spain. Nature 362:534–537.

Arsuaga, J. L., I. Martínez, A. Gracia, J. M. Carretero, C. Lorenzo, and N. García

1997   Sima de los Huesos (Sierra de Atapuerca, Spain). The Site. Journal of Human Evolution 33:109–127.

Arsuaga, J. L., I. Martínez, Ana Gracía, and C. Lorenzo

1997   The Sima de los Huesos Crania (Sierra de los Atapuerca, Spain). A Comparative Study. Journal of Human Evolution 33:219–281.

Asfaw, Berhane

1984   Preparation of the Bodo Cranium (abstract). American Journal of Physical Anthropology 63:135.

Azzaroli, A., M. Boccaletti, E. Delson, G. Moratti, and D. Torre

1986   Chronological and Paleogeographical Background to the Study of *Oreopithecus bambolii*. Journal of Human Evolution 15:533–540.

Bada, J. L., R. Gillespie, J. A. J. Gowlett, and R. E. M. Hedges

1984   Accelerator Mass Spectrometry Radiocarbon Ages of Amino Acid Extracts from Californian Palaeoindian Skeletons. Nature 312:442–444.

Bada, J. L., and P. M. Helfman

1975   Amino Acid Racemization Dating of Fossil Bones. World Archaeology 7:160–173.

Bada, J. L., R. A. Schroeder, and F. G. Carter

1974   New Evidence for the Antiquity of Man in North America Deduced from Aspartic Acid Racemization. Science 184:791–793.

Badam, G. L.

1989   Observations on the Fossil Hominid Site at Hathnora, Madhya Pradesh, India. In (A. Sahni and R. Gaur, editors) Perspectives in Human Evolution, pp. 153–172. Delhi: Renaissance Publishing House.

Badgley, Catherine, John Keiley, David Pilbeam, and Steven Ward

1984   The Paleobiology of South Asian Miocene Hominoids. In (John R. Lukacs, editor) The People of South Asia, pp. 3–28. New York: Plenum Press.

Bahn, Paul G.

1996   Treasure of the Sierra Atapuerca. Archaeology 49(1): 45–48.

Bartsiokas, A., and M. H. Day

1993   Lead Poisoning and Dental Caries in the Broken Hill Hominid. Journal of Human Evolution 24:243–249.

Bar Yosef, O., and B. Vandermeersch

1981   Notes Concerning the Possible Age of the Mousterian Layers in Qafzeh Cave. In (P. Sanlaville and J. Cauvin, editors) Prehistoire du Levant, pp. 281–285. Paris: Centre National de la Recherche Scientifique.

1993   Modern Humans in the Levant. Scientific American 268(4): 94–100.

Bass, William M.

1995   Human Osteology: A Laboratory and Field Manual of the Human Skeleton, fourth edition. Columbia, Missouri: Missouri Archaeological Society.

198

Beard, K. Christopher, Mark F. Teaford, and Alan Walker
    1986   New Wrist Bones of *Proconsul africanus* and *Proconsul nyanzae* from Rusinga Island, Kenya. Folia Primato-logica 47:97–118.

Begun, David R.
    1993   New Catarrhine Phalanges from Rudabanya (Northeastern Hungary) and the Problem of Parallelism and Convergence in Hominoid Postcranial Morphology. Journal of Human Evolution 24:373–402.
    1994a   Relations among the Great Apes and Humans: New Interpretations Based on the Fossil Great Ape *Dryo-pithecus.* Yearbook of Physical Anthropology 37:11–63.
    1994b   The Significance of *Otavipithecus namibiensis* to Interpretations of Hominoid Evolution. Journal of Human Evolution 27:385–394.

Begun, David R., Salvador Moyà-Solà, and M. Köhler
    1990   New Miocene Hominoid Specimens from Can Llobateres (Valles Penedes, Spain) and Their Geological and Paleoecological Context. Journal of Human Evolution 19:255–268.

Begun, David R., Mark F. Teaford, and Alan Walker
    1994   Comparative and Functional Anatomy of Proconsul Phalanges from the Kaswanga Primate Site, Rusinga Island. Journal of Human Evolution 26:98–166.

Behrensmeyer, Anna K.
    1984   Taphonomy and the Fossil Record. American Scientist 72:558–566.

Behrensmeyer, Anna K., and Andrew P. Hill
    1980   Fossils in the Making: Vertebrate Taphonomy and Paleoecology. Chicago: University of Chicago Press.

Benefit, Brenda R., and Monte L. McCrossin
    1995   Miocene Hominoids and Hominid Origins. Annual Review of Anthropology 24:237–256.

Berge, Christine, Rosine Orban-Segebarth, and Peter Schmid
    1984   Obstetrical Interpretation of the Australopithecine Pelvic Cavity. Journal of Human Evolution 13:573–587.

Bird, Junius
    1938   Antiquity and Migrations of the Early Inhabitants of Patagonia. Geographical Review 28:250–275.

Bischoff, J. L., J. A. Fitzpatrick, L. León, J. L. Arsuaga, C. Falgueres, J. J. Bahain, and T. Bullen
    1997   Geology and Preliminary Dating of the Hominid-Bearing Sedimentary Fill of the Sima de los Huesos Chamber, Cueva Mayor of the Sierra de Atapuerca, Burgos, Spain. Journal of Human Evolution 33:129–154.

Bischoff, James L., and Robert J. Rosenbauer
    1981   Uranium Series Dating of Human Skeletal Remains from the Del Mar and Sunnyvale Sites, California. Science 213:1003–1005.

Bishop, W. W.
    1964   More Fossil Primates and Other Miocene Mammals from Northeast Uganda. Nature 203:1327–1331.

Bishop, W. W., J. A. Miller, and F. J. Fitch
    1969   New Potassium-Argon Age Determinations Relevant to the Miocene Fossil Mammal Sequence in East Africa. American Journal of Science 267:669–699.

Black, Davidson (editor)
    1933   Fossil Man in China, the Choukoutien Cave Deposits with a Synopsis of Our Present Knowledge of the Late Cenozoic in China. Geological Survey of China Memoir 11 (series A).

Blanc, A.
    1939   L'Homme Fossile du Monte Circeo. L'Anthropologie 49: 253–264.

Boule, M.
    1911–13 L'Homme Fossile de La Chapelle-aux-Saints. Annals de Paléontologie 6:111–172: 7:21–56, 85–192; 8:1–70.

Bower, B.
    1987   Family Feud: Enter the "Black Skull." Science News 131:58–59.

Bown, Thomas M., Mary J. Kraus, Scott L. Wing, John G. Fleagle, Bruce H. Tiffany, Elwyn L. Simons, and Carl F. Vondra
    1982   The Fayum Primate Forest Revisited. Journal of Human Evolution 11:603–632.

Brace, C. Loring
    1964   The Fate of the "Classic" Neanderthals: Consideration of Hominid Catastrophism. Current Anthropology 5:3–43.
    1988   The Stages of Human Evolution: Human and Cultural Origins, third edition. Englewood Cliffs, NJ: Prentice-Hall.

Brace, C. L., P. E. Mahler, and R. B. Rosen
    1973   Tooth Measurements and the Rejection of the Taxon "*Homo habilis.*" Yearbook of Physical Anthropology 16:50–68.

Brace, C. Loring, and Ashley Montagu
  1977   Human Evolution, An Introduction to Biological Anthropology, second edition. New York: Macmillan Publishing Co.

Brace, C. Loring, Harry Nelson, Noel Korn, and Mary L. Brace
  1979   Atlas of Human Evolution, second edition. New York: Holt, Rinehart and Winston.

Brace, C. Loring, Shao Xiang-qing, and Zhang Zhen-biao
  1984   Prehistoric and Modern Tooth Size in China. In (Fred H. Smith and Frank Spencer, editors) The Origins of Modern Humans, pp. 485–516. New York: Alan R. Liss.

Brain, C. K. (editor)
  1993   Swartkrans: A Cave's Chronicle of Early Man. Transvaal Museum Monograph No. 8.

Brain, C. K., C. S. Churcher, J. D. Clark, F. E. Grine, P. Shipman, R. L. Susman, A. Turner, and V. Watson
  1988   New Evidence of Early Hominids, Their Culture and Environment from the Swartkrans Cave, South Africa. South African Journal of Science 84:828–835.

Bräuer, Günter
  1989   The Evolution of Modern Humans: A Comparison of the African and Non-African Evidence. In (Paul Mellars and Chris Stringer, editors) The Human Revolution: Behavioural and Biological Perspectives on the Origins of Modern Humans, pp. 123–154. Princeton: Princeton University Press.

Bromage, Timothy G.
  1985   Taung Facial Remodeling: A Growth and Development Study. In (Phillip V. Tobias, editor) Hominid Evolution: Past, Present and Future, pp. 239–245. New York: Alan R. Liss.

Brooks, Alison S., and Bernard Wood
  1990   The Chinese Side of the Story. Nature 344:288–289.

Broom, Robert
  1947   Discovery of a New Skull of the South African Ape-Man, Plesianthropus. Nature 159:672.
  1949   Another New Type of Fossil Ape-Man. Nature 163:57.
  1950   Finding the Missing Link. London: Watts and Co.

Broom, Robert, and J. T. Robinson
  1950   Ape or Man? Nature 166:843–844.
  1952   Swartkrans Ape-Man. Transvaal Museum Memoir 6.

Broom, Robert, John T. Robinson, and G. W. H. Schepers
  1950   Sterkfontein Ape-Man Plesianthropus. Transvaal Museum Memoir 4.

Brothwell, D. R.
  1961   The People of Mount Carmel: A Reconsideration of their Position in Human Evolution. Proceedings of the Prehistoric Society 27:155–159.

Brown, Frank, John Harris, Richard Leakey, and Alan Walker
  1985   Early Homo erectus skeleton from West Lake Turkana, Kenya. Nature 316:788–792.

Brown, Peter
  1987   Pleistocene Homogeneity and Holocene Size Reduction: The Australian Human Skeletal Evidence. Archaeology in Oceania 22:41–67.

Bermúdez de Castro, Jose María
  1993   The Atapuerca Dental Remains: New Evidence (1987–1991 Excavations) and Interpretations. Journal of Human Evolution 24:339–371.

Bermúdez de Castro, J. M., J. L. Arsuaga, E. Carbonell, A. Rosas, I. Martínez, and M. Mosquera
  1997   A Hominid from the Lower Pleistocene of Atapuerca, Spain: Possible Ancestor to Neandertals and Modern Humans. Science 276:1392–1395.

Bush, Michael E., C. Owen Lovejoy, Donald C. Johanson, and Yves Coppens
  1982   Hominid Carpal, Metacarpal, and Phalangeal Bones Recovered from the Hadar Formation: 1974–1977 Collections. American Journal of Physical Anthropology 57:651–677.

Busk, G.
  1865   On a Very Ancient Cranium from Gibraltar. Report of the British Association for the Advancement of Science (Bath 1864): 91–92.

Calcagno, James M.
  1989   Mechanisms of Human Dental Reduction: A Case Study from Post-Pleistocene Nubia. University of Kansas Publications in Anthropology 18.

Campbell, Bernard G., and James D. Loy
  1996   Humankind Emerging, seventh edition. New York: HarperCollins Publishers.

Camps, G., and G. Oliver (editors)
  1970   L'Homme de Cro-Magnon. Paris: Arts et Metier Graphiques.

Capitan, L., and D. Peyrony
1909   Deux Squelettes au Mileiu des Foyers de l'Époque Mousterienne. Comptes Rendus de l'Academie Inscriptions et Belles-Lettres 797–806.

Carlson, David S.
1976   Temporal Variation in Prehistoric Nubian Crania. American Journal of Physical Anthropology 45:467–484.

Carlson, David S., and Dennis P. Van Gerven
1977   Masticatory Function and Post-Pleistocene Evolution in Nubia. American Journal of Physical Anthropology 46: 495–506.
1979   Diffusion, Biological Determinism, and Biocultural Adaptation. American Anthropologist 81:561–580.

Carretero, José Miguel, Juan Luis Arsuaga, and Carlos Lorenzo
1997   Clavicles, Scapulae and Humeri from the Sima de los Huesos Site (Sierra de Atapuerca, Spain). Journal of Human Evolution 33:357–408.

Cartmill, Matt, William L. Hylander, and James Shafland
1987   Human Structure. Cambridge: Harvard University Press.

Chang, K. C.
1962   New Evidence on Fossil Man in China. Science 136:749–760.

Cherfas, Jeremy
1983   Trees Have Made Man Upright. New Scientist 97:172–178.

Chiu Chung-lang, Ku Yu-min, Chang Yin-yun, and Chang Sen-shui
1973   Peking Man Fossils and Cultural Remains Newly Discovered at Choukoutien. Vertebrata Palasiatica 11: 109–131.

Clark, G. A.
1988   Some Thoughts on the Black Skull: An Archaeologist's Assessment of WT-17000 (*A. boisei*) and Systematics in Human Paleontology. American Anthropologist 90:357–371.

Clark, J. D., J. de Heinzelin, K. D. Schick, W. K. Hart, T. D. White, G. WoldeGabriel, R. C. Walter, G. Suwa, B. Asfaw, E. Vrba, and Y. H. Selassie
1994   African *Homo erectus*: Old Radiometric Ages and Young Oldowan Assemblages in the Middle Awash Valley, Ethiopia. Science 264:1907–1910.

Clark, J. Desmond, Berhane Asfaw, Getaneh Assefa, J. W. K. Harris, H. Kurashina, R. C. Walter, T. D. White, and M. A. J. Williams
1984   Palaeoanthropological Discoveries in the Middle Awash Valley. Nature 307:423–428.

Clark, J. Desmond, Kenneth P. Oakley, L. H. Wells, and J. A. C. McClelland (contributors)
1947   New Studies on Rhodesian Man. Journal of the Royal Anthropological Institute 77:7–32.

Clarke, R. J.
1976   New Cranium of *Homo erectus* from Lake Ndutu, Tanzania. Nature 262:485–487.
1977   The Cranium of the Swartkrans Hominid, SK 847, and Its Relevance to Human Origins. Ph.D. dissertation, University of the Witwatersrand, Johannesburg.
1990   The Ndutu Cranium and the Origin of *Homo sapiens*. Journal of Human Evolution 19:699–736.
1994   On Some New Interpretations of Sterkfontein Stratigraphy. South African Journal of Science 90:211–214.

Clarke, R. J., and F. Clark Howell
1972   Affinities of the Swartkrans 847 Hominid Cranium. American Journal of Physical Anthropology 37:319–336.

Clarke, R. J., F. Clark Howell, and C. K. Brain
1970   More Evidence of an Advanced Hominid at Swartkrans. Nature 225:1219–1222.

Cohen, Perry
1996   Fitting a Face to Ngaloba. Journal of Human Evolution 30:373–379.

Collett, Margot
1933   A Study of Twelfth and Thirteenth Dynasty Skulls from Kerma (Nubia). Biometrika 25:254–283.

Condemi, S.
1985   Les Hommes Fossiles de Saccopastore (Italie) et Leurs Relations Phylogenetiques. Ph.D. dissertation, University of Bordeaux.

Conroy, Glenn C.
1987   The Taung Skull Revisited: New Evidence from High-Resolution Computed Tomography. AnthroQuest 38:10–11.
1990   Primate Evolution. New York: W. W. Norton & Co.
1994   *Otavipithecus*: How to Build a Better Hominid—Not. Journal of Human Evolution 27:373–383.
1997   Reconstructing Human Origins. New York: W. W. Norton & Co.

Conroy, Glenn C., Clifford J. Jolly, Douglas Cramer, and Jon E. Kalb
1978   Newly Discovered Fossil Hominid Skull from the Afar Depression, Ethiopia. Nature 276:67–70.

Conroy, Glenn C., Jeff W. Lichtman, and Lawrence B. Martin

1995   Brief Communication: Some Observations on Enamel Thickness and Enamel Prism Packing in the Miocene Hominoid *Otavipithecus namibiensis*. American Journal of Physical Anthropology 98:595–600.

Conroy, Glenn C., Martin Pickford, Brigette Senut, John Van Couvering, and Pierre Mein

1992   *Otavipithecus namibiensis*, First Miocene Hominoid from Southern Africa. Nature 356:144–148.

Conroy, Glenn C., and Michael W. Vannier

1987   Dental Development of the Taung Skull from Computerized Tomography. Nature 329:625–627.

1989   The Taung Skull Revisited: New Evidence from High-Resolution Computed Tomography. South African Journal of Science 85:30–32.

Conway, B., J. McNabb, and N. Ashton (editors)

1996   Excavations at Barnfield Pit, Swanscombe, 1968–72. British Museum Occasional Paper No. 94. London: The British Museum.

Cook, Della Collins, Jane E. Buikstra, C. Jean DeRousseau, and D. Carl Johanson

1983   Vertebral Pathology in the Afar Australopithecines. American Journal of Physical Anthropology 60:83–101.

Currey, John

1984   The Mechanical Adaptations of Bones. Princeton: Princeton University Press.

Dahlberg, Albert A. (editor)

1971   Dental Morphology and Evolution. Chicago: University of Chicago Press.

Dart, Raymond A.

1925   *Australopithecus africanus*: The Man-Ape of South Africa. Nature 115:195–199.

1929   A Note on the Taungs Skull. South African Journal of Science 26:648–458.

1954   The Second, or Adult, Female Mandible of *Australopithecus prometheus*. American Journal of Physical Anthropology 12:313–343.

1962   A Cleft Adult Mandible and the Nine other Lower Jaw Fragments from Makapansgat. American Journal of Physical Anthropology 20:267–286.

1967   Adventures with the Missing Link. Philadelphia: The Institutes Press.

Davis, P. R., and John Napier

1963   A Reconstruction of the Skull of *Proconsul africanus* (R.S. 51). Folia Primatologica 1:20–28.

Day, Michael H.

1969   Omo Human Skeletal Remains. Nature 222:1135–1138.

1986   Guide to Fossil Man: A Handbook of Human Palaeontology, fourth edition. Chicago: University of Chicago Press.

Day, M. H., M. D. Leakey, and C. Magori

1980   A New Hominid Fossil Skull (L.H. 18) from the Ngaloba Beds, Laetoli, Northern Tanzania. Nature 284:55–56.

Day, M. H., R. E. F. Leakey, A. C. Walker, and B. A. Wood

1976   New Hominids from East Turkana, Kenya. American Journal of Physical Anthropology 45:369–436.

Day, M. H., and J. R. Napier

1964   Hominid Fossils from Bed I, Olduvai Gorge, Tanganyika: Fossil Foot Bones. Nature 201:967–970.

Day, M. H., and B. A. Wood

1968   Functional Affinities of the Olduvai Hominid 8 Talus. Man 3:440–455.

Defleur, Alban, Olivier Dutour, Hélène Valladas, and Bernard Vandermeersch

1993   Cannibals among the Neanderthals? Nature 362:214.

Demes, Brigetta, and Norman Creel

1988   Bite Force, Diet, and Cranial Morphology of Fossil Hominids. Journal of Human Evolution 17:657–670.

Delson, Eric

1986a   An Anthropoid Enigma: Historical Introduction to the Study of *Oreopithecus bambolii*. Journal of Human Evolution 15: 523–540.

1986b   Human Phylogeny Revised Again. Nature 22:496–497.

Delson, Eric (editor)

1985   Ancestors: The Hard Evidence. New York: Alan R. Liss.

Dibble, Harold, and Michael Lenoir (editors)

1995   The Middle Paleolithic Site of Combe-Capelle Bas (France). Philadelphia: The University Museum, University of Pennsylvania.

Dijgenraam, Felix

1993   "Java Man" Gains (and Loses) a Consort. Science 261:297.

Dillehay, Tom D., and David J. Meltzer (editors)

1991   The First Americans: Search and Research. Boca Raton: CRC Press.

Dingwall, Doris, and Matthew Young
1933   The Skulls from Excavations at Dunstable, Bedfordshire. Biometrika 25:148–157.

Dubois, Eugene
1922   The Proto-Australian Fossil Man of Wadjak, Java. Koninklijke Akademie Wetenschappen te Amsterdam (series B) 23:1013–1051.

Ennouchi, E.
1962   Un Néandertalien: L'Homme der Jebel Irhoud (Maroc). L'Anthropologie 66:279–299.
1963   Les Néanderthaliens du Jebel Irhoud (Maroc). Comptes Rendus de l'Academie des Sciences de Paris 256:2459–2460.

Etler, Dennis A.
1984   The Fossil Hominoids of Lufeng, Yunnan Province, The People's Republic of China: A Series of Translations. Yearbook of Physical Anthropology 27:1–55.
1996   The Fossil Evidence for Human Evolution in Asia. Annual Review of Anthropology 25:275–301.

Falk, Dean
1980   A Reanalysis of the South African Australopithecine Natural Endocasts. American Journal of Physical Anthropology 53: 525–539.
1983a   Cerebral Cortices of East African Early Hominids. Science 221:1072–1074.
1983b   The Taung Endocast: A Reply to Holloway. American Journal of Physical Anthropology 60:479–489.
1983c   A Reconsideration of the Endocast of *Proconsul africanus*: Implications for Primate Brain Evolution. In (R. L. Ciochon and R. Corruccini, editors) New Interpretations of Ape and Human Ancestry, pp. 239–248. New York: Plenum Press.
1984   The Petrified Brain. Natural History 93:36–39.
1987   Hominid Paleoneurology. Annual Review of Anthropology 16:13–30.
1988   Enlarged Occipital/Marginal Sinuses and Emissary Foramina: Their Significance in Hominid Evolution. In (Frederick E. Grine, editor) Evolutionary History of the "Robust" Australopithecines, pp. 85–96. New York: Aldine de Gruyter.
1989   Ape-Like Endocast of "Ape-Man" Taung. American Journal of Physical Anthropology 80:335–339.

Falk, Dean, Charles Hildebolt, and Michael W. Vannier
1989   Reassessment of the Taung Early Hominid from a Neurological Perspective. Journal of Human Evolution 18:485–492.

Field, Henry
1948   Early Man in Mexico. Man 48:17–19.

Fleagle, J. G.
1983   Locomotor Adaptations of Oligocene and Miocene Hominoids and Their Phyletic Implications. In (R. L. Ciochon and R. Corruccini, editors) New Interpretations of Ape and Human Ancestry, pp. 301–324. New York: Plenum Press.
1988   Primate Adaptation and Evolution. San Diego: Academic Press.

Fleagle, J. G., T. M. Bown, J. D. Obradovich, and E. L. Simons
1986   How Old are the Fayum Primates? In (J. G. Else and P. C. Lee, editors) Primate Evolution, pp. 3–17. New York: Cambridge University Press.

Fleagle, J. G., and R. F. Kay
1983   New Interpretations of the Phyletic Position of Oligocene Hominoids. In (R. L. Ciochon and R. Corruccini, editors) New Interpretations of Ape and Human Ancestry, pp. 181–210. New York: Plenum Press.

Fleagle, J. G., and E. L. Simons
1982a   The Humerus of *Aegyptopithecus zeuxis*, a Primitive Anthropoid. American Journal of Physical Anthropology 59: 175–194.
1982b   Skeletal Remains of *Propliopithecus chirobates* from the Egyptian Oligocene. Folia Primatologica 39: 161–177.

Frankel, Victor H., and Margareta Nordin
1980   Basic Biomechanics of the Skeletal System. Philadelphia: Lea & Febiger.

Frayer, David W.
1973   *Gigantopithecus* and Its Relationship to *Australopithecus*. American Journal of Physical Anthropology 39:413–426.
1978   Evolution of the Dentition in Upper Paleolithic and Mesolithic Europe. University of Kansas Publications in Anthropology 10.

Frayer, David W., and Mary D. Russell
1987   Artificial Grooves on the Krapina Neanderthal Teeth. American Journal of Physical Anthropology 74:393–405.

Freedman, Leonard, and Mancel Lofgren
    1983   Human Skeletal Remains from Lake Tandou, New South Wales. Archaeology in Oceania 18:98–105.

Gambier, Dominique
    1989   Fossil Hominids from the Early Upper Palaeolithic (Aurignacian) of France. In (Paul Mellars and Chris Stringer, editors) The Human Revolution: Behavioural and Biological Perspectives on the Origins of Modern Humans, pp. 194–211. Princeton: Princeton University Press.

Gantt, G. D., N. I. Xirotiris, B. Kurten, and J. Melentis
    1980   The Petralona Dentition—Hominid or Cave Bear? Journal of Human Evolution 9:483–486.

Garrod, D. A. E., and D. M. A. Bate
    1937   The Stone Age of Mound Carmel I: Excavations at the Wadi el-Mughara. Oxford: Oxford University Press.

Gebo, D. L.
    1989   Locomotor and Phylogenetic Considerations in Anthropoid Evolution. Journal of Human Evolution 18:201–233.
    1992   Plantigrady and Foot Adaptation in African Apes: Implications for Hominid Origins. American Journal of Physical Anthropology 89:29–58.
    1993   Postcranial Anatomy and Locomotor Adaptation in Early African Anthropoids. In (Daniel L. Gebo, editor) Postcranial Adaptation in Nonhuman Primates, pp. 252–272. DeKalb: Northern Illinois University Press.
    1996   Climbing, Brachiation, and Terrestrial Quadrupedalism: Historical Precursors of Hominoid Bipedalism. American Journal of Physical Anthropology 101:55–92.

Gebo, Daniel L., K. Christopher Beard, Mark F. Teaford, Alan C. Walker, Susan Larson, William L. Jungers, and John G. Fleagle
    1988   A Hominoid Proximal Humerus from the Early Miocene of Rusinga Island, Kenya. Journal of Human Evolution 17:393–401.

Gebo, Daniel L., Laura MacLatchy, Robert Kityo, Alan Deino, John Kingston, and David Pilbeam
    1997   A Hominoid Genus from the Early Miocene of Uganda. Science 276:4010404.

Gebo, Daniel L., and Elwyn L. Simons
    1987   Morphology and Locomotor Adaptations of the Foot in Early Oligocene Anthropoids. American Journal of Physical Anthropology 74:83–101.

Gelvin, Bruce R.
    1980   Morphometric Affinities of *Gigantopithecus*. American Journal of Physical Anthropology 53:541–568.

Genovés, Santiago
    1960   Reevaluation of Age, Stature, and Sex of the Tepexpan Remains, Mexico. American Journal of Physical Anthropology 18:205–217.

Gervais, P.
    1872   Sur un Singe Fossile, d'Espèce non Encore Décite, qui a Été Découvert au Monte-Bambolii (Italie). Comptes Rendus de l'Academie des Sciences de Paris 74: 1217–1223.

Gibbons, Ann
    1996   *Homo erectus* in Java: A 250,000-Year Anachronism. Science 274:1841–1842.
    1997   Bone Sizes Trace the Decline of Man (and Woman). Science 276:896–897.

Goldberg, Kathy E.
    1982   The Skeleton: Fantastic Framework. Washington, DC: U.S. News Books.

Gorjanović-Kramberger, Dragutin
    1906   Der Diluviale Mensch von Krapina in Kroatien: Ein Beitrag zur Paläoanthropologie. Wiesbaden: Kreidel.

Gowlett, John
    1984   Ascent to Civilization: The Archaeology of Early Man. New York: Alfred A. Knopf.

Grayson, Donald K.
    1988   Americans Before Columbus: Perspectives on the Archaeology of the First Americans. In (Ronald C. Carlisle, editor) Americans Before Columbus: Ice-Age Origins, pp. 107–123. Ethnology Monographs 12, Department of Anthropology, University of Pittsburg.

Greene, David Lee, and George Armelagos
    1972   The Wadi Halfa Mesolithic Population. Department of Anthropology, University of Massachusetts, Research Reports 11.

Greene, David Lee, George H. Ewing, and George J. Armelagos
    1967   Dentition of a Mesolithic Population from Wadi Halfa, Sudan. American Journal of Physical Anthropology 27:41–56.

Greenfield, Leonard O.
    1990   Canine "Honing" in *Australopithecus afarensis*. American Journal of Physical Anthropology 82:135–143.

Gregory, William King
  1922   The Origin and Evolution of the Human Dentition. Baltimore: Williams and Wilkins.

Gregory, William King (editor)
  1950   The Anatomy of the Gorilla. New York: Columbia University Press.

Gregory, William K., and Milo Hellman
  1926   The Dentition of *Dryopithecus* and the Origin of Man. Anthropological Papers of the American Museum of Natural History 28, part 1.

Grine, Frederick E.
  1985   Dental Morphology and the Systematic Affinities of the Taung Fossil Hominid. In (Phillip V. Tobias, editor) Hominid Evolution: Past, Present and Future, pp. 247–253. New York: Alan R. Liss.

Grine, Frederick E. (editor)
  1988   Evolutionary History of the "Robust" Australopithecines. New York: Aldine de Gruyter.

Grine, F. E., B. Demes, W. L. Jungers, and T. M. Cole III
  1993   Taxonomic Affinity of the Early *Homo* Cranium from Swartkrans, South Africa. American Journal of Physical Anthropology 92:411–426.

Grün, Rainer
  1993   Electron Spin Resonance Dating in Paleoanthropology. Evolutionary Anthropology 2:172–181.
  1996   A Re-Analysis of Electron Spin Resonance Dating Results Associated with the Petrolona Cranium. Journal of Human Evolution 30:227–241.
  1997   ESR Analysis of Teeth from the Palaeoanthropological Site of Zhoukoudian, China. Journal of Human Evolution 32:89–91.

Grün, R., and C. B. Stringer
  1991   Electron Spin Resonance Data on the Evolution of Modern Humans. Archaeometry 33:153–199.

Grün, Rainer, and Alan Thorne
  1997   Dating the Ngandong Humans. Science 276:1575.

Hammond, Allen L.
  1983   Tales of an Elusive Ancestor. Science 83 4:36–44.

Hanihara, Kazuro, and Hiroshi Ueda
  1982   Dentition of the Minatogawa Man. In (H. Suzuki and K. Hanihara, editors) The Minatogawa Man, the Upper Pleistocene Man from the Island of Okinawa. The University Museum, University of Tokyo Bulletin 19, pp. 51–59.

Harris, John M.
  1985   Age and Paleoecology of the Upper Laetolil Beds, Laetoli, Tanzania. In (Eric Delson, editor) Ancestors: The Hard Evidence, pp. 76–81. New York: Alan R. Liss.

Harrison, Terry
  1986   A Reassessment of the Phylogenetic Relationships of *Oreopithecus bambolii* Gervais. Journal of Human Evolution 15: 541–584.
  1987   The Phylogenetic Relationships of the Early Catarrhine Primates: A Review of the Current Evidence. Journal of Human Evolution 16:41–80.

Harrison, Terry, and Lorenzo Rook
  1997   Enigmatic Anthropoid or Misunderstood Ape? The Phylogenetic Status of *Oreopithecus bambolii* Reconsidered. In (David R. Begun, Carol V. Ward, and Michael D. Rose, editors) Function, Phylogeny, and Fossils: Miocene Hominoid Evolution and Adaptations, pp. 327–362. New York: Plenum Press.

Häusler, M., and P. Schmid
  1995   Comparison of the Sts 14 and AL 288-1: Implications for Birth and Sexual Dimorphism in Australopithecines. Journal of Human Evolution 29:363–383.

Hedges, R E. M., R. A. Housley, C. R. Bronk, and G. J. Van Klinken
  1992   Radiocarbon Dates from the Oxford AMS System: *Archaeometry* Datelist 15. Archaeometry 34:337–357.

Heim, J. L.
  1976   Les Hommes Fossiles de La Ferrassie. Archives de l'Institut Paléontologie Humaine Mémoire 35:1–331.
  1982   Les Enfants Neandertaliens de La Ferrassie. Paris: Masson.

Heizer, Robert F., and Sherburne F. Cook
  1959   New Evidence of Antiquity of Tepexpan and Other Human Remains from the Valley of Mexico. South-Western Journal of Anthropology 15:36–42.

Hennig, G. J., W. Herr, E. Weber, and N. I. Xirotiris
  1981   ESR-Dating of the Fossil Hominid Cranium from Petralona Cave, Greece. Nature 292:533–536.

Hewes, G. W., H. T. Irwin, M. Papworth, and A. Saxe
  1964   A New Fossil Human Population from the Wadi Halfa Area. Nature 203:341–343.

205

Higgs, E. S., and D. R. Brothwell
1961   North Africa and Mount Carmel: Recent Developments. Man 61:138–139.

Hildebrand, Milton, Dennis M. Bramble, Karel F. Liem, and David B. Wake (editors)
1985   Functional Vertebrate Morphology. Cambridge: Harvard University Press.

Hill, Andrew, and Steven Ward
1988   Origin of the Hominidae: The Record of African Large Hominoid Evolution between 14 My and 4 My. Yearbook of Physical Anthropology 31:49–83.

Hillson, Simon
1996   Dental Anthropology. Cambridge: Cambridge University Press.

Holloway, Ralph L.
1973   Endocranial Volumes of Early African Hominids, and the Role of the Brain in Human Mosaic Evolution. Journal of Human Evolution 2:449–459.
1975   The Role of Human Social Behavior in the Evolution of the Brain. Forty-Third James Arthur Lecture on the Evolution of the Human Brain. New York: The American Museum of Natural History.
1981a   The Indonesian *Homo erectus* Brain Endocasts Revisited. American Journal of Physical Anthropology 55:503–521.
1981b   Volumetric and Asymmetry Determinations on Recent Hominid Endocasts: Spy I and II, Djebel Irhoud I, and the Salè *Homo erectus* Specimens, with Some Notes on Neandertal Brain Size. American Journal of Physical Anthropology 55:385–393.
1984   The Taung Endocast and the Lunate Sulcus: A Rejection of the Hypothesis of its Anterior Position. American Journal of Physical Anthropology 64:285–287.
1988   "Robust" Australopithecine Brain Endocasts: Some Preliminary Observations. In (Frederick E. Grine, editor) Evolutionary History of the "Robust" Australopithecines, pp. 97–105. New York: Aldine de Gruyter.

Howell, F. Clark
1951   The Place of Neanderthal Man in Human Evolution. American Journal of Physical Anthropology 9:379–416.
1958   Upper Pleistocene Men of the Southwest Asian Mousterian. In (G. H. R. von Koenigswald, editor) Hundert Jahre Neanderthaler (Neanderthal Centenary), pp. 185–198. Utrecht: Kemink en Zoon N.V.
1960   European and Northwest African Middle Pleistocene Hominids. Current Anthropology 1:195–232.
1970   Early Man. New York: Time-Life Books.
1978   Hominidae. In (Vincent J. Maglio and H. B. S. Cooke, editors) Evolution of African Mammals, pp. 154–258. Cambridge: Harvard University Press.

Howells, W. W.
1977   Paleoanthropology in the People's Republic of China. Washington: National Academy of Sciences.
1985   Taung: A Mirror for American Anthropology. In (Phillip V. Tobias, editor) Hominid Evolution: Past, Present and Future, pp. 19–24. New York: Alan R. Liss.

Hrdlicka, Ales
1924   New Data on the Teeth of Early Man and Certain European Anthropoid Apes. American Journal of Physical Anthropology 7:109–137.
1930   The Skeletal Remains of Early Man. Smithsonian Miscellaneous Collections No. 83.
1937   The "Minnesota Man." American Journal of Physical Anthropology 22:175–199.

Hublin, J. J.
1994   Pathological Aspects of the Middle Pleistocene Skull of Salé (Morocco). American Journal of Physical Anthropology, Supplement 18:110.

Hughes, Alun R., and Phillip V. Tobias
1977   A Fossil Skull Probably of the Genus *Homo* from Sterkfontein, Transvaal. Nature 265:310–312.

Hürzeler, J.
1958   *Oreopithecus bambolii* Gervais: A Preliminary Report. Verh. Naturf. Ges. Basel 69:1–48.
1960   The Significance of *Oreopithecus* in the Genealogy of Man. Triangle 4:164–174.

Hutchinson, Dale L., Clark Spencer Larsen, and Inui Choi
1997   Stressed to the Max? Physiological Perturbation in the Krapina Neandertals. Current Anthropology 38:904–914.

Huxley, Thomas Henry
1863   Man's Place in Nature. 1959 reprinted edition, University of Michigan Press, Ann Arbor.

Ikeya, Motoji
1982   Matters Arising: Petralona Cave Dating Controversy. Nature 299:281.

Institute of Vertebrate Paleontology and Paleoanthropology (Chinese Academy of Sciences)
1980   Atlas of Primitive Man in China. Beijing: Science Press.

Jacob, Teuku
1975a   L'Homme de Java. La Recherche 6:1027–1032.

1975b   Morphology and Paleoecology of Early Man in Java. In (R. H. Tuttle, editor) Paleoanthropology, Morphology, and Paleoecology, pp. 311–325. The Hauge: Mouton.

Jaeger, J. J.
1975a   The Mammalian Faunas and Hominid Fossils of the Middle Pleistocene of the Maghreb. In (Karl W. Butzer and Glynn Ll. Isaac, editors) After the Australopithecines, pp. 399–418. The Hague: Mouton.
1975b   Découverte d'un Crâne d'Hominidé dans le Pléistocène Moyen du Maroc. Centre National de la Recherche Scientifique, Colloques Internationaux, Problèmes Actuels de Paléontologie (Evolution des Vertébrés) 218:897–902.
1981   Les Hommes Fossiles du Pléistocene Moyen du Maghreb dans leur Cadre Géologique, Chronologique, et Paléoécologique. In (Becky A. Sigmon and Jerome S. Cybulski, editors) *Homo erectus*: Papers in Honor of Davidson Black, pp. 159–188. Toronto: University of Toronto Press.

James, Jamie
1989   Stalking the Giant Ape. Discover 10:42–50.

Jantz, R. L., and Douglas W. Owsley
1997   Pathology, Taphonomy, and Cranial Morphometrics of the Spirit Cave Mummy. Nevada Historical Society Quarterly 40:62–84.

Jelinek, Arthur J.
1982   The Tabun Cave and Paleolithic Man in the Levant. Science 216:1369–1375.

Jenks, Albert Ernest
1937   Minnesota's Browns Valley Man and Associated Burial Artifacts. Memoirs of the American Anthropological Association No. 49.

Jia Lanpo
1980   Early Man in China. Beijing: Foreign Languages Press.

Johanson, Donald C.
1976   Ethiopia Yields First "Family" of Early Man. National Geographic 150:789–811.
1979   A Consideration of the "*Dryopithecus* Pattern." Ossa 6: 125–137.

Johanson, Donald, and Blake Edward
1996   From Lucy to Language. New York: Simon & Schuster Editions.

Johanson, Donald C., and Maitland A. Edey
1981   Lucy, the Beginnings of Humankind. New York: Simon and Schuster.

Johanson, Donald C., C. Owen Lovejoy, William H. Kimbel, Tim D. White, Steven C. Ward, Michael E. Bush, Bruce M. Latimer, and Yves Coppens
1982   Morphology of the Pliocene Partial Hominid Skeleton (A.L. 288-1) from the Hadar Formation, Ethiopia. American Journal of Physical Anthropology 57:403–451.

Johanson, Donald C., Fidelis T. Masao, Gerald G. Eck, Tim D. White, Robert C. Walter, William H. Kimbel, Berhane Asfaw, Paul Manega, Prosper Ndessokia, and Gen Suwa
1987   New Partial Skeleton of *Homo habilis* from Olduvai Gorge, Tanzania. Nature 327:205–209.

Johanson, Donald, and James Shreeve
1989   Lucy's Child: The Discovery of a Human Ancestor. New York: Early Man Publishing.

Johanson, D. C., and M. Taieb
1976   Plio-Pleistocene Hominid Discoveries in Hadar, Ethiopia. Nature 260:293–297.

Johanson, D. C., and T. D. White
1979   A Systematic Assessment of Early Atrican Hominids. Science 203:321–330.
1986   Fossil Debate. Discover 7:116.

Johanson, Donald C., Tim D. White, and Yves Coppens
1978   A New Species of the Genus *Australopithecus* (Primates: Hominidae) from the Pliocene of Eastern Africa. Kirtlandia 28.
1982   Dental Remains from the Hadar Formation, Ethiopia: 1974–1977. American Journal of Physical Anthropology 57: 545–603.

Jungers, William L.
1982   Lucy's Limbs: Skeletal Allometry and Locomotion in *Australopithecus afarensis*. Nature 297:676–678.
1987   Body Size and Morphometric Affinities of the Appendicular Skeleton in *Oreopithecus bambolii* (IGF 11778). Journal of Human Evolution 16:445–456.
1988   Lucy's Length: Stature Reconstruction in *Australopithecus afarensis* (A.L. 288-1) with Implications for Other Small Bodied Hominids. American Journal of Physical Anthropology 76:227–231.

Jungers, William L., and Jack T. Stern, Jr.
1983   Body Proportions, Skeletal Allometry and Locomotion in the Hadar Hominids: A Reply to Wolpoff. Journal of Human Evolution 12:673–684.

Kalb, Jon E., Clifford J. Jolly, Elizabeth B. Oswald, and Paul F. Whitehead
1984   Early Hominid Habitation in Ethiopia. American Scientist 72:168–178.

Kay, Richard F.
1982   *Sivapithecus simonsi*, a New Species of Miocene Hominoid with Comments on the Phylogenetic Status of the Ramapithecinae. International Journal of Primatology 3:113–174.

Kay, Richard F., John G. Fleagle, and Elwyn L. Simons
1981   A Revision of the Oligocene Apes from the Fayum Province, Egypt. American Journal of Physical Anthropology 55: 293–322.

Kay, Richard F., and Elwyn L. Simons
1980   The Ecology of Oligocene African Anthropoidea. International Journal of Primatology 1:22–37.
1983   A Reassessment of the Relationship between Late Miocene and Subsequent Hominoidea. In (R. L. Ciochon and R. Corruccini, editors) New Interpretations of Ape and Human Ancestry, pp. 577–624. New York: Plenum Press.

Keith, Arthur
1911   The Early History of the Gibraltar Cranium. Nature 87:314.

Keith, Arthur, and T. D. McCown
1937   Mount Carmel Man: His Bearing on the Ancestry of Modern Races. In (G. G. MacCurdy, editor) Early Man, pp. 44–52. Philadelphia: Lippincott.

Kelley, Jay
1986   Species Recognition and Sexual Dimorphism in *Proconsul* and *Rangwapithecus*. Journal of Human Evolution 15:461–495.

Kelley, Jay, and Dennis Etler
1989   Hominoid Dental Variability and Species Number at the Late Miocene Site of Lufeng, China. American Journal of Primatology 18: 15–34.

Kelley, Jay, and David Pilbeam
1986   The Dryopithecines: Taxonomy, Comparative Anatomy and Phylogeny of Miocene Large Hominoids. In (D. R. Swindler and J. Erwin, editors) Comparative Primate Biology, vol. 1: Systematics, Evolution, and Anatomy, pp. 361–411. New York: Alan R. Liss.

Kennedy, Kenneth A. R.
1959   The Aboriginal Population of the Great Basin. University of California Archaeological Survey Reports 45: 1–84.
1975   Neanderthal Man. Minneapolis: Burgess Publishing Co.
1994   Evolution of South Asian Pleistocene Hominids: Demic Displacement or Regional Continuity? In (A. Parpola and P. Koskikallio, editors) South Asian Archaeology 1993, pp. 337–344. Helsinki: Suomalainen Tiedeakatemia.

Kennedy, Kenneth A. R., Arun Sonakia, John Chiment, and K. K. Verma
1991   Is the Narmada hominid an Indian *Homo erectus*? American Journal of Physical Anthropology 86:475–496.

Kidd, R. S., P. O'Higgins, and C. E. Oxnard
1996   The OH8 Foot: A Reappraisal of the Functional Morphology of the Hindfoot Utilizing a Multivariate Analysis. Journal of Human Evolution 31:269–291.

Kimbel, W. H.
1988   Identification of a Partial Cranium of *Australopithecus afarensis* from the Koobi Fora Formation, Kenya. Journal of Human Evolution 17:647–656.

Kimbel, William H., Donald C. Johanson, and Yves Coppens
1982   Pliocene Hominid Cranial Remains from the Hadar Formation, Ethiopia. American Journal of Physical Anthropology 57:453–499.

Kimbel, William H., Donald C. Johanson, and Yoel Rak
1994   The First Skull and Other New Discoveries of *Australopithecus afarensis* at Hadar, Ethiopia. Nature 368:449–451.
1996   Systematic Assessment of a Maxilla of *Homo* from Hadar, Ethiopia. American Journal of Physical Anthropology 103:235–262.

Kimbel, W. H., R. C. Walter, D. C. Johanson, K. M. Reed, J. C. Aronson, Z. Assefa, C. W. Marean, G. G. Eck, R. Bobe, E. Hovers, Y. Rak, C. Vondra, T. Yemane, D. York, Y. Chen, N. M. Evensen, and P. E. Smith
1996   Late Pliocene *Homo* and Oldowan Tools from the Hadar Formation (Kada Hadar Member), Ethiopia. Journal of Human Evolution 31:549–561.

Kimbel, William H., and Tim D. White
1988   A Revised Reconstruction of the Adult Skull of *Australopithecus afarensis*. Journal of Human Evolution 17:545–550.

Kimbel, William H., Tim D. White, and Donald C. Johanson

1984   Cranial Morphology of *Australopithecus afarensis*: A Comparative Study Based on a Composite Reconstruction of the Adult Cranium. American Journal of Physical Anthropology 64:337–388.

1988   Implications of KNM-WT 17000 for the Evolution of "Robust" *Australopithecus*. In (Frederick E. Grine, editor) Evolutionary History of the "Robust" *Australopithecines*, pp. 259–268. New York: Aldine de Gruyter.

King, William

1864   The Reputed Fossil Man of the Neanderthal. Quarterly Journal of Science 1:88–97.

Klaatsch, H., and O. Hauser

1909   *Homo aurignacensis* hauseri. Prähistorisches Zeitschrift 1:273–338.

Klein, Richard G.

1973   Geological Antiquity of Rhodesian Man. Nature 244:311–312.

1988   The Causes of "Robust" Australopithecine Extinction. In (Frederick E. Grine, editor) Evolutionary History of the "Robust" Australopithecines, pp. 499–505. New York: Aldine de Gruyter.

1989   The Human Career: Human Biological and Cultural Origins. Chicago: University of Chicago Press.

Koenigswald, G. H. R. von

1952   *Gigantopithecus blacki* von Koenigswald, a Giant Fossil Hominoid from the Pleistocene of Southern China. Anthropological Papers of the American Museum of Natural History 43, part 4.

1956   Meeting Prehistoric Man. London: Thames and Hudson.

Koenigswald, G. H. R. von, and Franz Weidenreich

1939   The Relationship between *Pithecanthropus* and *Sinanthropus*. Nature 144:926–929.

Koppel, R.

1935   Das Alter der Neuentdeckten Schädel von Nazareth. Biblica 16:58–73.

Kordos, László, and David R. Begun

1977   A New Reconstruction of RUD 77, a Partial Cranium of *Dryopithecus brancoi* from Rudabánya, Hungary. American Journal of Physical Anthropology 103:277–294.

Koritzer, Richard T., and Lucile St. Hoyme

1977   Dental Pathology of the Rhodesian Man. Journal of the American Dental Association 99:642–643.

1980   Caries and Elemental Composition of the Rhodesian Man Dentition. Journal of the Washington Academy of Sciences 70:74–79.

Kraatz, Reinhart

1985a   A Review of Recent Research on Heidelberg Man, *Homo erectus heidelbergensis*. In (Eric Delson, editor) Ancestors: The Hard Evidence, pp. 268–271. New York: Alan R. Liss.

1985b   Recent Research on Heidelberg Jaw, *Homo erectus heidelbergensis*. In (Phillip V. Tobias, editor) Hominid Evolution: Past, Present and Future, pp. 313–318. New York: Alan R. Liss.

Krings, Matthias, Anne Stone, Ralf W. Schmitz, Heike Krainitzki, Stoneking, and Svante Pääbo

1997   Neandertal DNA Sequences and the Origin of Modern Humans. Cell 90:1930.

Kuman, K.

1994   The Archaeology of Sterkfontein—Past and Present. Journal of Human Evolution 27:471–495.

Lamy, Paul

1983   Le Systeme Podal de Certains Hominides Fossiles du Plio-Pleistocène d'Afrique de l'Ést: Étude Morpho-Dynamique. L'Anthropologie 87:435–464.

Langdon, John H.

1986   Functional Morphology of the Miocene Hominoid Foot. In (F. S. Szalay, editor) Contributions to Primatology 22. Zurich: Karger.

Larsen, Clark Spencer

1982   The Anthropology of St. Catherines Island: 3. Prehistoric Human Biological Adaptation. Anthropological Papers of the American Museum of Natural History 57, part 3.

1984   Health and Disease in Prehistoric Georgia: The Transition to Agriculture. In (Mark Nathan Cohen and George J. Armelagos, editors) Paleopathology at the Origins of Agriculture, pp. 367–392. Orlando: Academic Press.

1985b   Dental Modifications and Tool Use in the Western Great Basin. American Journal of Physical Anthropology 67:393–402.

1995   Biological Changes in Human Populations with Agriculture. Annual Review of Anthropology 24:185–213.

1997   Bioarchaeology: Interpreting Behavior from the Human Skeleton. Cambridge: Cambridge University Press.

Larsen, Clark Spencer (editor)

1985a   The Antiquity and Origin of Native North Americans. New York: Garland Publishing.

1990   The Archaeology of Mission Santa Catalina de Guale: 2. Biocultural Interpretations of a Population in Transition. Anthropological Papers of the American Museum of Natural History No. 68.

Larsen, Clark Spencer, and Robert L. Kelly
   1995   Bioarchaeology of the Stillwater Marsh: Prehistoric Human Adaptation in the Western Great Basin. Anthropological Papers of the American Museum of Natural History No. 77.

Larsen, Clark Spencer, and Robert M. Matter
   1985   Human Origins: The Fossil Record. Prospect Heights, IL: Waveland Press.

Larsen, Clark Spencer, Robert M. Matter, and Daniel L. Gebo
   1991   Human Origins: The Fossil Record, second edition. Prospect Heights, IL: Waveland Press.

Larsen, Clark Spencer, and Thomas C. Patterson
   1997   Paleoanthropology: Americas. In (Frank Spencer, editor) History of Physical Anthropology: An Encyclopedia, Volume One, pp. 68–74. New York: Garland Publishing.

Larsen, Clark Spencer, Margaret J. Schoeninger, Dale L. Hutchinson, Katherine F. Russell, and Christopher B. Ruff
   1990   Beyond Demographic Collapse: Biological Adaptation and Change in Native Populations of La Florida. In (David Hurst Thomas, editor) Columbian Consequences, vol. 2: Archaeological and Historical Perspectives on the Spanish Borderlands East, pp. 409–428. Washington: Smithsonian Institution Press.

Lartet, E.
   1856   Note sur un Grand Singe Fossile qui se Rattache au Groupe des Singes Supérieurs. Comptes Rendus de L'Academie Sciences de Paris 43:219–228.

Lasker, Gabriel W., and Robert N. Tyzzer
   1981   Physical Anthropology, third edition. New York: Holt, Rinehart and Winston.

Latimer, Bruce, and C. Owen Lovejoy
   1989   The Calcaneus of *Australopithecus afarensis* and Its Implications for the Evolution of Bipedality. American Journal of Physical Anthropology 78:369–386.
   1990   Hallucal Tarsometatarsal Joint in *Australopithecus afarensis*. American Journal of Physical Anthropology 82:125–133.

Leakey, L. S. B.
   1960   Finding the World's Earliest Man. National Geographic 118:420–435.
   1961   New Finds at Olduvai Gorge. Nature 189:649–650.
   1974   By the Evidence: Memoirs, 1932–1951. New York: Harcourt, Brace and Jovanovich.

Leakey, L. S. B., and M. D. Leakey
   1964   Recent Discoveries of Fossil Hominids in Tanganyika: At Olduvai and Near Lake Natron. Nature 202:5–7.

Leakey, L. S. B., P. V. Tobias, and J. R. Napier
   1964   A New Species of the Genus *Homo* from Olduvai Gorge. Nature 202:7–9.

Leakey, M. D.
   1971   Olduvai Gorge, vol. 3: Excavations in Beds I and II, 1960–1963. Cambridge: Cambridge University Press.
   1984   Disclosing the Past: An Autobiography. Garden City: Doubleday and Co.

Leakey, M. D., R. J. Clarke, and L. S. B. Leakey
   1971   New Hominid Skull from Bed I, Olduvai Gorge, Tanzania. Nature 232:308–312.

Leakey, M. D., and J. M. Harris
   1987   Laetoli: A Pliocene Site in Northern Tanzania. New York: Oxford University Press.

Leakey, M. D., R. L. Hay, G. H. Curtis, R. E. Drake, M. K. Jackes, and T. D. White
   1976   Fossil Hominids from the Laetolil Beds. Nature 262:460–466.

Leakey, Meave G., Craig S. Feibel, Ian McDougall, and Alan Walker
   1995   New Four-Million-Year-Old Hominid Species from Kanapoi and Allia Bay, Kenya. Nature 376:565–571.

Leakey, Richard E. F.
   1969   Early *Homo sapiens* Remains from the Omo River Region of Southwest Ethiopia. Nature 222:1137–1138.
   1970   In Search of Man's Past at Lake Rudolf. National Geographic 137:712–733.
   1971   Further Evidence of Lower Pleistocene Hominids from East Rudolf, North Kenya. Nature 321:241–245.
   1973a   Skull 1470. National Geographic 143:818–829.
   1973b   Further Evidence of Lower Pleistocene Hominids from East Rudolf, North Kenya. Nature 242:170–173.
   1973c   Evidence for an Advanced Plio-Pleistocene Hominid from East Rudolf, Kenya. Nature 242:447–450.
   1974   Further Evidence of Lower Pleistocene Hominids from East Rudolf, North Kenya. Nature 248:653–656.
   1976   Hominids in Africa. American Scientist 64: 174–178.
   1981   The Making of Mankind. New York: Dutton.

Leakey, R. E., and M. G. Leakey
   1986a   A New Miocene Hominoid from Kenya. Nature 324:143–146.
   1986b   A Second New Miocene Hominoid from Kenya. Nature 324:146–148.

Leakey, Richard E. F., Meave G. Leakey, and Anna K. Behrensmeyer
  1978   The Hominid Catalogue. In (Meave G. Leakey and Richard E. Leakey, editors) Koobi Fora Research Project, vol. 1: The Fossil Hominids and an Introduction to Their Context, 1968–1974, pp. 86–187. Oxford: Clarendon Press.

Leakey, Richard E., Meave G. Leakey, and Alan C. Walker
  1988a   Morphology of *Turkanapithecus kalakolensis* from Kenya. American Journal of Physical Anthropology 76:277–288.
  1988b   Morphology of *Afropithecus turkanensis* from Kenya. American Journal of Physical Anthropology 76:289–307.

Leakey, Richard E., and Roger Lewin
  1977   Origins. New York: E. P. Dutton.

Leakey, Richard E. F., J. M. Mungai, and Alan C. Walker
  1971   New Australopithecines from East Rudolf, Kenya. American Journal of Physical Anthropology 35:175–186.

Leakey, Richard E. F., and Alan C. Walker
  1976   *Australopithecus*, *Homo erectus* and the Single Species Hypothesis. Nature 261:572–574.
  1985a   Further Hominids from the Plio-Pleistocene of Koobi Fora, Kenya. American Journal of Physical Anthropology 67: 135–163.
  1985b   *Homo erectus* Unearthed. National Geographic 168:624–629.
  1988   New *Australopithecus boisei* Specimens from East and West Lake Turkana, Kenya. American Journal of Physical Anthropology 76:1–24.
  1989   Early *Homo erectus* from West Lake Turkana, Kenya. In (Giacomo Giacobini, editor) Hominidae, Proceedings of the 2nd International Congress of Human Paleontology, pp. 209–215. Milan, Italy: Jaca Book.

Le Gros Clark, W. E.
  1928   Rhodesian Man. Man 28:206–207.
  1971   The Antecedents of Man, third edition. Chicago: Quadrangle Books.

Le Gros Clark, W. E., and L. S. B. Leakey
  1951   Fossil Mammals of Africa, No. 1. The Miocene Hominoidea of East Africa. London: British Museum (Natural History).

Lewin, Roger
  1982   Fossil Lucy Grows Younger, Again. Science 219:43–44.
  1983a   Were Lucy's Feet Made for Walking? Science 220:700–702.
  1983b   Do Ape-Size Legs Mean Ape-Like Gait? Science 221:537–538.
  1984   Ancestors Worshiped. Science 224:477–479.
  1985   Surprise Findings in the Taung Child's Face. Science 228:42–44.
  1986   New Fossil Upsets Human Family. Science 233:720–721.
  1987   Bones of Contention: Controversies in the Search for Human Origins. New York: Simon and Schuster.

Liritzis, Y.
  1982   Matters Arising: Petralona Cave Dating Controversy. Nature 299:280–281.

Liu Ze-Chun
  1983   Le Remplissage de la Grotte de l'Homme de Pekin Choukoutien—Localite 1. L'Anthropologie 87:163–176.

Lovejoy, C. Owen
  1974   The Gait of Australopithecines. Yearbook of Physical Anthropology 17:174–161.
  1975   Biomechanical Perspectives on the Lower Limb of Early Hominids. In (R. H. Tuttle, editor) Primate Morphology and Evolution, pp. 291–326. The Hague: Mouton.
  1976   The Locomotor Skeleton of Basal Pleistocene Hominids. Colloque VI: Les Plus Anciens Hominides. IX Congres, Union Internationale des Sciences Préhistoriques et Protohistoriques, Nice.
  1978   A Biomechanical Review of the Locomotor Diversity of Early Hominids. In (C. J. Jolly, editor) Early Hominids of Africa, pp. 403–429. New York: St. Martin's.
  1981   The Origin of Man. Science 211:341–350.
  1984   The Natural Detective. Natural History 93:24–28.
  1988   Evolution of Human Walking. Scientific American 256:118–125.

Lu Qingwu, Xu Qinghua, and Zheng Liang
  1981   Preliminary Research on the Cranium of *Sivapithecus yunnanensis*. Vertebrata Palasiatica 19:101–106. (Translation in Yearbook of Physical Anthropology 27:23–27.)

Lumley, Marie-Antoinette de
  1981   Les Anteneandertaliens en Europe. In (Becky A. Sigmon and Jerome S. Cybulski, editors) *Homo erectus*: Papers in Honor of Davidson Black, pp. 115–132. Toronto: University of Toronto Press.

Lumley, Henry de, and Marie-Antoinette de Lumley
1973   Pre-Neanderthal Human Remains from Arago Cave in Southeastern France. Yearbook of Physical Anthropology 17:162–168.

Lynch, Thomas F.
1990   Glacial-Age Man in South America? A Critical Review. American Antiquity 55:12–36.

MacLatchy, Laura M., and William H. Bossert
1996   An Analysis of the Articular Distribution of the Femoral Head and Acetabulum in Anthropoids, with Implications for Hip Function in Miocene Hominoids. Journal of Human Evolution 31:425–454.

Magori, C. C.
1980   Laetoli Hominid 18: Studies on a Pleistocene Fossil Human Skull from Northern Tanzania. Ph.D. dissertation, University of London.

Magori, C. C., and Michael H. Day
1983   An Early *Homo sapiens* Skull from the Ngaloba Beds, Laetoli, Northern Tanzania. Anthropos 10:143–183.

Maier, Wolfgang O., and Abel T. Nkini
1984   Olduvai Hominid 9: New Results of Investigation. In (Peter Andrews and Jens Lorenz Franzen, editors) The Early Evolution of Man, with Special Emphasis on Southeast Asia and Africa, pp. 123–130. Courier Forschungsinstitut Senckenberg 69.
1985   The Phylogenetic Position of the Olduvai Hominid 9, Especially as Determined from Basicranial Evidence. In (Eric Delson, editor) Ancestors: The Hard Evidence, pp. 249–254. New York: Alan R. Liss.

Mann, Alan
1975   Paleodemographic Aspects of the South African Australopithecines. University of Pennsylvania Publications in Anthropology 1.
1981   The Significance of the *Sinanthropus* Casts and Some Paleodemographic Notes. In (Becky A. Sigmon and Jerome S. Cybulski, editors) *Homo erectus*: Papers in Honor of Davidson Black, pp. 41–62. Toronto: University of Toronto Press.
1988   The Nature of Taung Dental Maturation. Nature 333:123.

Mann, Alan, and Erik Trinkaus
1973   Neandertal and Neandertal-like Fossils from the Upper Pleistocene. Yearbook of Physical Anthropology 17:169–193.

Marshall, Eliot
1990   Paleoanthropology Gets Physical. Science 247:798–801.

Marston, A. T.
1936   Preliminary Note on a New Fossil Human Skull from Swanscombe. Nature 138:200–201.
1937   The Swanscombe Skull. Journal of the Royal Anthropological Institute 67:339–406.

Martin, Debra L., George J. Armelagos, Alan H. Goodman, and Dennis P. Van Gerven
1984   The Effects of Socioeconomic Change in Prehistoric Nubia. In (Mark Nathan Cohen and George J. Armelagos, editors) Paleopathology at the Origins of Agriculture, pp. 193–214. Orlando: Academic Press.

Martin, H.
1923   L'Homme Fossile de La Quina. Archives de Morphologie Génerale et Experimental 15:1–253.

Martin, Lawrence
1986   Relationships among Extant and Extinct Great Apes and Humans. In (B. Wood, L. Martin, and P. Andrews, editors) Major Topics in Primate and Human Evolution, pp. 161–187. Cambridge: Cambridge University Press.

Martínez, J., and J. L. Arsuaga
1997   The Temporal Bones from Sima de los Huesos Middle Pleistocene Site (Sierra de Atapuerca, Spain). A Phylogenetic Approach. Journal of Human Evolution 33:283–318.

Massone, Mauricio
1996   Hombre Temprano y Paleoambiente en la Región de Magallanes: Evaluación Crítica y Perspectivas. Annales Instituto Patagonia 24:81–98.

Matiegka, J.
1934   *Homo predmostensis*: Fosilni Clovek z Predmosti na Morave. Prague: Ceská Academie Ved a Umeni.
1938   *Homo predmostensis*. II. Autres Partes du Squelette. Prague: Ceska Academie Ved a Umeni.

McCollum, Melanie A., and Steven C. Ward
1997   Subnasoalveolar Anatomy and Hominoid Phylogeny: Evidence from Comparative Ontogeny. American Journal of Physical Anthropology 102: 377–405.

McCown, T. D., and Arthur Keith
1939   The Stone Age of Mount Carmel II: The Fossil Human Remains from the Levalloiso-Mousterian. Oxford: Clarendon Press.

McCrossin, Monte L., and Brenda R. Benefit
    1993   Recently Recovered *Kenyapithecus* Mandible and Its Implications for Great Ape and Human Origins. Proceedings of the National Academy of Sciences 90:1962–1966.

McHenry, H. M.
    1984   Size and Shape of the Hominoid Foot: Implications for Function and Evolution. American Journal of Physical Anthropology 63:192.
    1986   The First Bipeds: A Comparison of the *A. afarensis* and *A. africanus* Postcranium and Implications for the Evolution of Bipedalism. Journal of Human Evolution 15:177–191.

McHenry, Henry M., Peter Andrews, and Robert S. Corruccini
    1980   Miocene Hominoid Palatofacial Morphology. Folia Primatologica 33:241–252.

McKee, J. K.
    1993   Faunal Dating of the Taung Hominid Fossil Deposit. Journal of Human Evolution 25:363–376.

McKee, J. K., and P. V. Tobias
    1994   Taung Stratigraphy and Taphonomy: Preliminary Results based on the 1988–93 Excavations. South African Journal of Science 90:233–235.

McLean, Franklin C., and Marshall R. Urist
    1968   Bone: Fundamentals of the Physiology of Skeletal Tissue. Chicago: University of Chicago Press.

Mellars, Paul
    1989   Major Issues in the Emergence of Modern Humans. Current Anthropology 30:349–385.

Meltzer, David J.
    1995   Clocking the First Americans. Annual Review of Anthropology 24:21–45.

Mercier, N., H. Valladas, O. Bar-Yosef, B. Vandermeersch, C. Stringer, and J. L. Joron
    1993   Thermoluminescence Date for the Mousterian Burial Site of Es-Skhul, Mt. Carmel. Journal of Archaeological Science 20:169–174.

Merrilees, D.
    1973   Fossiliferous Deposits at Lake Tandou, New South Wales, Australia. Memoirs of the National Museum of Victoria 34:177–182.

Molnar, Stephen, and I. M. Molnar
    1985   The Incidence of Enamel Hypoplasia among the Krapina Neandertals. American Anthropologist 87:536–549.

Morbeck, M. E.
    1983   Miocene Hominoid Discoveries from Rudabanya: Implications from the Postcranial Skeleton. In (R. L. Ciochon and R. Corruccini, editors) New Interpretations of Ape and Human Ancestry, pp. 369–404. New York: Plenum Press.

Moss, Melvin L.
    1960   A Re-evaluation of the Dental Status and Chronological Age of the Tepexpan Remains. American Journal of Physical Anthropology 18:71–72.

Moyà Soyà, Salvador, and Melke Köhler
    1993   Recent Discoveries of *Dryopithecus* Shed New Light on Evolution of Great Apes. Nature 365:543–545.

Mturi, A. A.
    1976   New Hominid from Lake Ndutu, Tanzania. Nature 262:484–485.

Murrill, Rupert Ivan
    1981   Petralona Man, a Descriptive and Comparative Study, with New Important Information on Rhodesian Man. Springfield: Charles C Thomas.

Napier, John
    1967   The Antiquity of Human Walking. Scientific American 216:56–66.

Napier, J. R., and P. R. Davis
    1959   The Forelimb Skeleton and Associated Remains of *Proconsul africanus*. Fossil Mammals of Africa, British Museum (Natural History) 16:1–69.

Ninkovich, D., and L. H. Burckle
    1978   Absolute Age of the Base of the Hominid-Bearing Beds in Eastern Java. Nature 275:306–308.

Oakley, Kenneth P.
    1952   Swanscombe Man. Proceedings of the Geological Association of London 63:271–300.

Oakley, Kenneth Page, Bernard Grant Campbell, and Theya Ivitsky Molleson
    1971   Catalogue of Fossil Hominids, Part II: Europe. London: British Museum (Natural History).
    1975   Catalogue of Fossil Hominids, Part III: Americas, Asia, Australasia. London: British Museum (Natural History).
    1977   Catalogue of Fossil Hominids, Part I: Africa, second edition. London: British Museum (Natural History).

Ogilvie, Marsha D., Bryan K. Curran, and Erik Trinkaus
1989   Incidence and Patterning of Dental Enamel Hypoplasia among the Neandertals. American Journal of Physical Anthropology 79:25–41.

Olson, Storrs L., and D. Tab Rasmussen
1986   Paleoenvironment of the Earliest Hominoids: New Evidence from the Oligocene Avifauna of Egypt. Science 233:1202–1204.

Oppenoorth, W. F. F.
1937   The Place of *Homo soloensis* among Fossil Men. In (G. G. MacCurdy, editor) Early Man, pp. 349–360. Philadelphia: Lippincott.

Ovey, C. D. (editor)
1964   The Swanscombe Skull. Occasional Papers of the Royal Anthropological Institute 20.

Owen, Roger C.
1984   The Americas: The Case Against an Ice-Age Human Population. In (Fred H. Smith and Frank Spencer, editors) The Origins of Modern Humans: A World Survey of the Fossil Evidence, pp. 517–563. New York: Alan R. Liss.

Oxnard, Charles E., and F. Peter Lisowski
1980   Functional Articulation of some Hominoid Foot Bones: Implications for the Olduvai (Hominid 8) Foot. American Journal of Physical Anthropology 42:107–117.

Pares, J. M., and A. Perez-Gonzalez
1995   Paleomagnetic Age for Hominid Fossils at Atapuerca Archaeological Site, Spain. Nature 269:830–832.

Patterson, Thomas Carl, and Junius B. Bird
1975   Chile. In (Kenneth P. Oakley, Bernard G. Campbell, and Theya I. Molleson, editors) Catalogue of Fossil Hominids, Part III: Americas, Asia, Australia, p. 12. London: British Museum (Natural History).

Pérez, Pilar-Julia, Ana Gracia, Ignacio Martínez, and Juan Luis Arsuaga
1997   Paleopathological Evidence of the Cranial Remains of the Sima de los Huesos Middle Pleistocene Site (Sierra de Atapuerca, Spain). Description and Preliminary Inferences. Journal of Human Evolution 33:409–421.

Peyer, Bernhard
1968   Comparative Odontology. Chicago: University of Chicago Press.

Pfeiffer, John
1984   Early Man Stages a Summit Meeting in New York City. Smithsonian 15(5): 50–57.
1985   The Emergence of Humankind, fourth edition. New York: Harper & Row.

Pilbeam, David
1970   *Gigantopithecus* and the Origins of the Hominidae. Nature 225:516–519.
1982   New Hominoid Skull Material from the Miocene of Pakistan. Nature 295:232–234.
1984a   Reflections on Early Human Ancestors. Journal of Anthropological Research 40:14–22.
1984b   The Descent of Hominoids and Hominids. Scientific American 250:84–96.
1989   Human Origins and Evolution. In (A. C. Fabian, editor) Origins, pp. 89–114. New York: Cambridge University Press.
1996   Genetic and Morphological Records of the Hominoidea and Hominid Origins: A Synthesis. Molecular Phylogenetics and Evolution 5:155–168.

Pilbeam, David R., Michael D. Rose, John C. Barry, and S. M. Ibrahim Shah
1990   New *Sivapithecus humeri* from Pakistan and the Relationship of *Sivapithecus* and *Pongo*. Nature 348:237–239.

Pinkley, C.
1935–36 The Significance of Wadjak Man: A Fossil *Homo sapiens* from Java. Peking Natural History Bulletin 10:183–200.

Poirier, Frank E.
1993   Understanding Human Evolution, third edition. Englewood Cliffs, NJ: Prentice-Hall.

Pope, Geoffrey G.
1988   Recent Advances in Far Eastern Paleoanthropology. Annual Review of Anthropology 17:43–77.

Poulianos, Aris N.
1971   Petralona, a Middle Pleistocene Cave in Greece. Archaeology 24:6–11.
1982   Petralona Cave Dating Controversy. Nature 299:280.
1989   Petralona Cave within Lower-Middle Pleistocene Sites. Palaeogeography, Palaeoclimatology, and Palaeoecology 73:287–294.

Preuss, T. M.
1982   The Face of *Sivapithecus indicus*: Description of a New, Relatively Complete Specimen from the Siwaliks of Pakistan. Folia Primatologica 38: 141–157.

214

Protsch, Reiner R. R.
1978   Catalog of Fossil Hominids of North America. New York: Gustav Fischer.

Pycraft, W. P. (and others)
1928   Rhodesian Man and Associated Remains. London: British Museum (Natural History).

Radinsky, Leonard B.
1973   *Aegyptopithecus* Endocasts: Oldest Record of a Pongid Brain. American Journal of Physical Anthropology 39:239–248.

Radovčić, Jakov
1985   Neanderthals and Their Contemporaries. In (Eric Delson, editor) Ancestors: The Hard Evidence, pp. 310–318. New York: Alan R. Liss.

Radovčić, Jakov, Fred H. Smith, Erik Trinkaus, and Milford H. Wolpoff
1988   The Krapina Hominids: An Illustrated Catalog of Skeletal Collection. Zagreb: Mladost and Croatian Natural History Museum.

Rak, Yoel
1983   The Australopithecine Face. New York: Academic Press.
1990   On the Differences between Two Pelvises of Mousterian Context from the Qafzeh and Kebara Caves, Israel. American Journal of Physical Anthropology 81:323–332.

Reader, John
1981   Missing Links. London: Collins.

Reed, Charles A.
1983   A Short History of the Discovery and Early Study of the Australopithecines: The First Find to the Death of Robert Broom (1924–1951). In (Kathleen J. Reichs, editor) Hominid Origins, Inquiries Past and Present, pp. 1–77. Washington, DC: University Press of America.

Reisner, George A.
1923   Excavations at Kerma, I-V. Harvard African Studies 5, 6.

Rensberger, Boyce
1984   Bones of Our Ancestors. Science 84 5:28–39.

Retallack, G. J., Erick A. Bestland, and Daniel P. Dugas
1995   Miocene paleosols and habitats of Proconsul on Rusinga Island, Kenya. Journal of Human Evolution 29:53–91.

Rightmire, G. Philip
1976   Relationships of Middle and Upper Pleistocene Hominids from Sub-Saharan Africa. Nature 260:238–240.
1979   Cranial Remains of *Homo erectus* from Beds 11 and IV, Olduvai Gorge, Tanzania. American Journal of Physical Anthropology 51:99–116.
1981   *Homo erectus* at Olduvai Gorge, Tanzania. In (Becky A. Sigmon and Jerome S. Cybulski, editors) *Homo erectus*: Papers in Honor of Davidson Black, pp. 189–192. Toronto: University of Toronto Press.
1983   The Lake Ndutu Cranium and Early *Homo sapiens* in Africa. American Journal of Physical Anthropology 61:245–254.
1984   *Homo sapiens* in Sub-Saharan Africa. In (Fred H. Smith and Frank Spencer, editors) The Origins of Modern Humans: A World Survey of the Fossil Evidence, pp. 295–325. New York: Alan R. Liss.
1990   The Evolution of *Homo erectus*: Comparative Anatomical Studies of an Extinct Human Species. New York: Cambridge University Press.
1996   The Human Cranium from Bodo, Ethiopia: Evidence for Speciation in the Middle Pleistocene. Journal of Human Evolution 31:21–39.

Rink, W. J., H. P. Schwarcz, F. H. Smith, and J. Radovčić
1995   ESR Dates for Krapina Hominids. Nature 378:24.

Roberts, Richard G., Rhys Jones, and M. A. Smith
1990   Thermoluminescence Dating of a 50,000-year-old Human Occupation Site in Northern Australia. Nature 345: 153–156.

Robinson, J. T.
1953   The Nature of *Telanthropus capensis*. Nature 171:33.
1954   The Genera and Species of the Australopithecinae. Transvaal Museum Memoir 4.
1956   The Dentition of the Australopithecinae. Transvaal Museum Memoir 9.
1960   The Affinities of the New Olduvai Australopithecine. Nature 186:456–458.
1962   Australopithecines and the Origin of Man. The Smithsonian Report for 1961, pp. 479–500.
1965   *Homo "habilis"* and the Australopithecines. Nature 205:121–124.

Rogers, Spencer L.
1963   The Physical Characteristics of the Aboriginal La Jollan Population of Southern California. San Diego Museum Papers 4.

1974   An Ancient Human Skeleton Found at Del Mar, California. San Diego Museum Papers 7.

Rook, L., T. Harrison, and B. Engesser
1996   The Taxonomic Status and Biochronological Implications of New Finds of *Oreopithecus* from Baccinello (Tuscany, Italy). Journal of Human Evolution 30:3–27.

Rosas, Antonio
1997   A Gradient of Size and Shape for the Atapuerca Sample and Middle Pleistocene Variability. Journal of Human Evolution 33:319–331.

Rose, Michael D.
1983   Miocene Hominoid Postcranial Morphology: Monkey-Like, Ape-Like, Neither, or Both? In (R. L. Chiochon and R. Corruccini, editors) New Interpretations of Ape and Human Ancestry, pp. 405–420. New York: Plenum Press.
1986   Further Hominoid Postcranial Specimens from the Middle Miocene Chinji Formation, Pakistan. Journal of Human Evolution 15: 333–368.
1989   New Postcranial Specimens of Catarrhines from the Middle Miocene Chinji Formation, Pakistan: Descriptions and a Discussion of Proximal Humeral Functional Morphology in Anthropoids. Journal of Human Evolution 18:131–162.
1993   Locomotor Anatomy of Miocene Hominoids. In (Daniel L. Gebo, editor) Postcranial Adaptation in Nonhuman Primates, pp. 252–272. DeKalb: Northern Illinois University Press.

Rose, Mike D., Meave G. Leakey, Richard E. F. Leakey, and Alan Walker
1992   Postcranial Specimens of *Simiolus enjiessi* and Other Primitive Catarrhines from the Early Miocene of Lake Turkana, Kenya. Journal of Human Evolution 22:171–237.

Ruff, Christopher
1988   Hindlimb Articular Surface Allometry in Hominoidea and Macaca, with Comparisons to Diaphyseal Scaling. Journal of Human Evolution 17:687–714.
1991   Climate and Body Shape in Hominid Evolution. Journal of Human Evolution 21:81–106.
1993   Climatic Adaptation and Hominid Evolution: The Thermoregulatory Imperative. Evolutionary Anthropology 2:53–60.
1994   Morphological Adaptation to Climate in Modern and Fossil Hominids. Yearbook of Physical Anthropology 37:65–107.

Ruff, Christopher B., Erik Trinkaus, and Trenton W. Holliday
1997   Body Mass and Encephalization in Pleistocene *Homo*. Nature 387:173–176.

Ruff, Christopher B., Erik Trinkaus, Alan Walker, and Clark Spencer Larsen
1993   Postcranial Robusticity in *Homo*. I: Temporal Trends and Mechanical Interpretation. American Journal of Physical Anthropology 91:21–53.

Ruff, Christopher B., Alan Walker, and Mark F. Teaford
1989   Body Mass, Sexual Dimorphism and Femoral Proportions of Proconsul from Rusinga and Mfangano Islands, Kenya. Journal of Human Evolution 18:515–536.

Sanders, William J., and Brian E. Bodenbender
1994   Morphometric Analysis of Lumbar Vertebra UMP 67-28: Implications for Spinal Function and Phylogeny of the Miocene Moroto Hominoid. Journal of Human Evolution 26:203–237.

Santa Luca, A. P.
1980   The Ngandong Fossil Hominids: A Comparative Study of a Far Eastern *Homo erectus* Group. Yale University Publications in Anthropology 78.

Sarmiento, Esteban E.
1987   The Phyletic Position of *Oreopithecus* and Its Significance in the Origin of the Hominoidea. American Museum Novitates 2881:1–44.

Sartono, S.
1971   Observations on a New Skull of *Pithecanthropus erectus* (Pithecanthropus VIII) from Sangiran, Central Java. Proceedings of the Koninklijke Nederlandsche Akademie van Wetenschappen, Amsterdam, series B. 74:185–194.
1972   Discovery of Another Hominid Skull at Sangiran, Central Java. Current Anthropology 13:124–126.
1975   Implications Arising from Pithecanthropus VIII. In (R. H. Tuttle, editor) Paleoanthropology, Morphology and Paleoecology, pp. 327–360. The Hague: Mouton.

Sartono, S., and Dominique Grimaud-Hervé
1983   Les Pariétaux des Pithécanthropes Sangiran 12 et Sangiran 17. L'Anthropologie 87:475–482.

Sauer, Norman J., and Terrill W. Phenice
1977   Hominid Fossils: An Illustrated Key, second edition. Dubuque, IA: Wm. C. Brown Co.

Saxe, Arthur A.
1971   Social Dimensions of Mesolithic Practices in a Mesolithic Population from Wadi Halfa, Sudan. Society for American Archaeology, Memoir 25:39–57.

Schepartz, Lynne Alison
1987   From Hunters to Herders: Subsistence Pattern and Morphological Change in Eastern Africa. Ph.D. dissertation, University of Michigan, Ann Arbor.

Schoetensack, O.
1908   Der Unterkiefer des *Homo heidelbergensis* aus den Sanden von Mauer bei Heidelberg. Leipzig: Wilhelm Englemann.

Schultz, A. H.
1969   The Life of Primates. London: Weidenfeld and Nicolson.

Schwarcz, H. P., R. Grün, B. Vandermeersch, O. Bar-Yosef, H. Valladas, and E. Tchernov
1988   ESR Dates for the Hominid Burial Site of Qafzeh in Israel. Journal of Human Evolution 17:733–737.

Schwartz, Jeffrey, H.
1984a   The Evolutionary Relationships of Man and the Orangs. Nature 308:501–505.
1984b   Hominoid Evolution: A Review and a Reassessment. Current Anthropology 25:655–672.
1995   Skeleton Keys: An Introduction to Human Skeletal Morphology, Development, and Analysis. New York: Oxford University Press.

Sergi, S.
1931   Le Crane Neanderthalien de Saccopastore. L'Anthropologie 41:241–247.
1939   Der Neandertalschadel vom Monte Circeo. Anthropologischer Anzeiger 16:203–217.
1948   The Palaeanthropi in Italy: The Fossil Men of Saccopastore and Circeo. Man 48:61–64, 76–79. (Reprinted in Adam, or Ape: a Sourcebook of Discoveries about Early Man, edited by L. S. B. Leakey, J. Prost, and S. Prost, pp. 229–237. Cambridge, MA: Schenkman Publishing Co.)
1958   Die Neandertalischen Palaeanthropen in Italien. In (G. H. R. von Koenigswald, editor) Hundert Jahre Neanderthaler, pp. 38–51. Utrecht: Kemink en Zoon N.V.
1962   Morphological Position of the *"Prophaneranthropi"* (Swanscombe and Fontechevade). In (William Howells, editor) Ideas on Human Evolution: Selected Essays, 1949–1961, pp. 507–520. Cambridge: Harvard University Press.

Shapiro, Harry L.
1974   Peking Man: The Discovery, Disappearance and Mystery of a Priceless Scientific Treasure. New York: Simon and Schuster.

Shipman, Pat
1981   Life History of a Fossil: An Introduction to Taphonomy and Paleoecology. Cambridge: Harvard University Press.
1986   Baffling Limb on Family Tree. Discover 7:86–93.

Shipman, Pat, Alan Walker, and David Bichell
1985   The Human Skeleton. Cambridge: Harvard University Press.

Shreeve, James
1994   "Lucy," Crucial Early Human Ancestor, Finally Gets a Head. Science 264:34–35.
1996   New Skeleton Gives Path from Trees to Ground an Odd Turn. Science 272:654.

Simons, Elwyn L.
1967   The Earliest Apes. Scientific American 217:28–35.
1972   Primate Evolution: An Introduction to Man's Place in Nature. New York: Macmillan.
1987   New Faces of *Aegyptopithecus* from the Oligocene of Egypt. Journal of Human Evolution 16:273–290.
1989   Human Origins. Science 245:1343–1350.
1995   Egyptian Oligocene Primates: A Review. Yearbook of Physical Anthropology 38:199–238.

Simons, E. L., and S. Chopra
1969   A New Species of *Gigantopithecus* (Hominoidea, Primates) from North India with Some Comments on Its Relationship to Earliest Hominids. Postilla 138:1–18.

Simons, Elwyn L., and Peter C. Ettel
1970   *Gigantopithecus*. Scientific American 222:76–84.

Simons, E. L., and D. Pilbeam
1965   Preliminary Revision of the Dryopithecinae. Folia Primatologica 3:81–152.

Simpson, Scott W.
1996   *Australopithecus afarensis* and Human Evolution. In (Carol R. Ember, Melvin Ember, and Peter Peregrine, editors) Research Frontiers in Anthropology, pp. 3–28. Needham: Simon & Schuster Custom Publishing.

Skelton, Randall R., and Henry M. McHenry
1992   Evolutionary Relationships among Early Hominids. Journal of Human Evolution 23:309–350.

Skinner, Mark F., and Geoffrey H. Sperber
1982   Atlas of Radiographs of Early Man. New York: Alan R. Liss.

Smith, B. Holly
1986   Dental Development in *Australopithecus* and Early *Homo*. Nature 323:327–330.
1990   KNM-WT 15000 and the Life History of *Homo erectus*. American Journal of Physical Anthropology 81:296.

Smith, Fred H.
1976a   The Neandertal Remains from Krapina, Northern Yugoslavia: An Inventory of the Upper Limb Remains. Zeitschrift für Morphologie und Anthropologie 67:275–290.
1976b   The Neandertal Remains from Krapina: A Descriptive and Comparative Study. Report of Investigations, Department of Anthropology, University of Tennessee, Knoxville 15.
1976c   The Skeletal Remains of the Earliest Americans: A Survey. Tennessee Anthropologist 1:116–147.
1977   On the Application of Morphological "Dating" to the Hominid Fossil Record. Journal of Anthropological Research 33:302–316.
1980   Sexual Differences in European Neanderthal Crania with Special Reference to the Krapina Remains. Journal of Human Evolution 9:359–375.
1984   Fossil Hominids from the Upper Pleistocene of Central Europe and the Origin of Modern Europeans. In (Fred H. Smith and Frank Spencer, editors) The Origins of Modern Humans: A World Survey of the Fossil Evidence, pp. 137–209. New York: Alan R. Liss.

Smith, Fred H., Anthony B. Falsetti, and Steven M. Donnelly
1989   Modern Human Origins. Yearbook of Physical Anthropology 32:35–68.

Smith, Fred H., and Maria O. Smith
1986   On the Significance of Anomalous Nasal Bones in the Neandertals from Krapina. In (Vladimir V. Novotny and Alena Mizerova, editors) Fossil Man: New Facts—New Ideas, pp. 217–226. Brno: Anthropos Institute-Moravian Museum. (Special issue of Anthropos 23)

Snow, Charles E.
1953   The Ancient Palestinian: Skhul V Reconstruction. American School of Prehistoric Research Bulletin 17:5–10.

Solecki, Ralph S.
1957   Shanidar Cave. Scientific American 197:58–64.
1963   Prehistory in Shanidar Valley, Northern Iraq. Science 139:179–193.
1971   Shanidar, the First Flower People. New York: Alfred A. Knopf.

Sonakia, Arun
1985   Early *Homo* from Narmada Valley, India. In (Eric Delson, editor) Ancestors: The Hard Evidence, pp. 334–338. New York: Alan R. Liss.

Spear, C. Fred, Paul Y. Sondaar, and S. Tassier Hussain
1991   A New Hominoid Hamate and First Metacarpal from the Late Miocene Nagri Formation of Pakistan. Journal of Human Evolution 21:413–424.

Steele, D. Gentry, and Claude A. Bramblett
1988   The Anatomy and Biology of the Human Skeleton. College Station: Texas A&M University Press.

Steele, D. Gentry, and Joseph F. Powell
1992   Peopling of the Americas: Paleobiological Evidence. Human Biology 64:303–336.
1993   Paleobiology of the First Americans. Evolutionary Anthropology 2:138–146.

Stern, Jack T., Jr., and Randall L. Susman
1983   The Locomotor Anatomy of *Australopithecus afarensis*. American Journal of Physical Anthropology 60:279–317.

Stewart, T. Dale
1958   First Views of the Shanidar I Skull. Sumer 14:90–96 (Reprinted in Annual Report of the Smithsonian Institution for 1958, pp. 473–480).
1961   The Skull of Shanidar I. Sumer 17:97–106 (Reprinted in Annual Report of the Smithsonian Institution for 1961, pp. 521–533).
1977   The Neanderthal Skeletal Remains from Shanidar Cave, Iraq: A Summary of the Findings to Date. Proceedings of the American Philosophical Society 121:121–165.
1985   Preliminary Report on an Early Human Burial in the Wadi Kubbaniya, Egypt. In (Phillip V. Tobias, editor) Hominid Evolution: Past, Present and Future, pp. 335–340. New York: Alan R. Liss.

Stewart, T. D., and Michael Tiffany
1986   Description of the Human Skeleton. In (Fred Wendorf and Romuald Schild) The Wadi Kubbanlya Skeleton: A Late Paleolithic Burial from Southern Egypt, pp. 49–53. Dallas: Southern Methodist University Press.

Stiner, Mary C.
1991   The Cultural Significance of Grotta Guattari Reconsidered: I. The Faunal Remains from Grotta Guattari: A Taphonomic Perspective. Current Anthropology 32:103–117.
1994   Honor among Thieves: A Zooarchaeological Study of Neandertal Ecology. Princeton: Princeton University Press.

Straus, William L., Jr.
1963   The Classification of *Oreopithecus*. In (S. L. Washburn, editor) Classification and Human Evolution, pp. 146–177. Chicago: Aldine.

Straus, William L., Jr., and A. J. E. Cave
1957   Pathology and the Posture of Neanderthal Man. Quarterly Review of Biology 32:348–363.

Stringer, Christopher B.
1974   A Multivariate Study of the Petralona Skull. Journal of Human Evolution 3:397–404.
1988   The Dates of Eden. Nature 331:565–566.
1993   Secrets of the Pit of the Bones. Nature 362:501–502.

Stringer, Christopher B., R. Grün, H. P. Schwarcz, and P. Goldberg
1989   ESR Dates for the Hominid Burial Site of Es Skhul in Israel. Nature 338:756–758.

Stringer, Christopher B., F. Clark Howell, and John K. Melentis
1979   The Significance of the Fossil Hominid Skull from Petralona, Greece. Journal of Archaeological Science 6:235–253.

Stringer, C. B., J. J. Hublin, and B. Vandermeersch
1984   The Origin of Anatomically Modern Humans in Western Europe. In (Fred H. Smith and Frank Spencer, editors) The Origins of Modern Humans, pp. 51–135. New York: Alan R. Liss.

Susman, Randall L.
1983   Evolution of the Human Foot: Evidence from Plio-Pleistocene Hominids. Foot Ankle 3:365–376.
1988   New Postcranial Remains from Swartkrans and Their Bearing on the Functional Morphology and Behavior of *Paranthropus robustus*. In (Frederick E. Grine, editor) Evolutionary History of the "Robust" Australopithecines, pp. 149–173. New York: Aldine de Gruyter.
1994   Fossil Evidence for Early Hominid Tool Use. Science 265:1570–1573.

Susman, Randall L., and Jack T. Stern
1982   Functional Morphology of *Homo habilis*. Science 217:931–934.

Susman, Randall L., Jack T. Stern, and William L. Jungers
1984   Arboreality and Bipedality in the Hadar Hominids. Folia Primatologica 43:113–156.
1985   Locomotor Adaptations in the Hadar Hominids. In (Eric Delson, editor) Ancestors: The Hard Evidence, pp. 184–192. New York: Alan R. Liss.

Suwa, Gen, Tim D. White, and F. Clark Howell
1996   Mandibular Postcanine Dentition from the Shungura Formation, Ethiopia: Crown Morphology, Taxonomic Allocations, and Plio-Pleistocene Hominid Evolution. American Journal of Physical Anthropology 101:247–282.

Suzuki, Hisashi
1982   Skull of the Minatogawa Man. In (H. Suzuki and K. Hanihara, editors) The Minatogawa Man, the Upper Pleistocene Man from the Island of Okinawa. The University Museum, University of Tokyo Bulletin 19:7–49.

Suzuki, Hisashi, and Kazuro Hanihara (editors)
1982   The Minatogawa Man, the Upper Pleistocene Man from the Island of Okinawa. The University Museum, University of Tokyo, Bulletin 19.

Suzuki, H., and F. Takai (editors)
1970   The Amud Man and His Cave Site. Tokyo: University of Tokyo Press.

Suzuki, Hisashi, and Giichi Tanabe
1982   Introduction. In (H. Suzuki and K. Hanihara, editors) The Minatogawa Man, the Upper Pleistocene Man from the Island of Okinawa. The University Museum, University of Tokyo, Bulletin 19: 1–6.

Swindler, Daris R.
1976   Dentition of Living Primates. London: Academic Press.

Swindler, Daris R., and Charles D. Wood
1973   An Atlas of Primate Gross Anatomy: Baboon, Chimpanzee, and Man. Seattle: University of Washington Press.

Swisher, C. C., III, W. J. Rink, S. C. Antón, H. P. Schwarcz, G. H. Curtis, A. Suprijo, and Widiasmoro
1996   Latest *Homo erectus* of Java: Potential Contemporaneity with *Homo sapiens* in Southeast Asia. Science 274:1870–1874.

Szalay, Frederick S., and Annalisa Berzi
1973   Cranial Anatomy of *Oreopithecus*. Science 180:183–185.

Szalay, Frederick S., and Eric Delson
1979   Evolutionary History of the Primates. New York: Academic Press.

Szalay, Frederick S., and John H. Langdon
1986   The Foot of *Oreopithecus*: An Evolutionary Assessment. Journal of Human Evolution 15:585–621.

Tague, Robert G., and C. Owen Lovejoy
1986   The Obstetric Pelvis of A.L. 288-1 (Lucy). Journal of Human Evolution 15:237–255.

Tappen, N. C.
1985   The Dentition of the "Old Man" of La Chapelle-aux-Saints and Inferences Concerning Neandertal Behavior. American Journal of Physical Anthropology 67:43–50.

Tattersall, Ian, and Eric Delson
1984   Ancestors: Four Million Years of Humanity. New York: American Museum of Natural History.

Tattersall, Ian, and G. J. Sawyer
1996   The Skull of "*Sinanthropus*" from Zhoukoudian, China: A New Reconstruction. Journal of Human Evolution 31:311–314

Taylor, R. E., and Louis A. Payen
1979   The Role of Archaeometry in American Archaeology: Approaches to the Evaluation of the Antiquity of *Homo sapiens* in California. Advances in Archaeological Method and Theory 2:239–283.

Taylor, R. E., L. A. Payen, C. A. Prior, P. J. Slota, Jr., R. Gillispie, J. A. J. Gowlett, R. E. M. Hedges, A. J. T. Jull, T. H. Zabel, D. J. Donahue, and R. Berger
1985   Major Revisions in the Pleistocene Age Assignments for North American Human Skeletons by C-14 Accelerator Mass Spectrometry: None Older than 11,000 C-14 Years B.P. American Antiquity 50:136–140.

Teaford, Mark F., K. Christopher Beard, Richard E. Leakey, and Alan Walker
1988   New Hominoid Facial Skeleton from the Early Miocene of Rusinga Island, Kenya, and Its Bearing on the Relationship between *Proconsul nyanzae* and *Proconsul africanus*. Journal of Human Evolution 17:461–477.

Terra, Hellmut de
1941   Pleistocene Formations and Stone Age Man in China. Peking: Institut de Geo-Biologie.
1947   Preliminary Notes on the Discovery of Fossil Man at Tepexpan in the Valley of Mexico. American Antiquity 13:40–44.

Terra, Hellmut de, Javier Romero, and T. D. Stewart
1949   Tepexpan Man. Viking Fund Publications in Anthropology 11.

Thomas, David Hurst
1987   The Archaeology of Mission Santa Catalina de Guale: 1. Search and Discovery. Anthropological Papers of the American Museum of Natural History 63, part 2.

Thorne, Alan G.
1971   Mungo and Kow Swamp: Morphological Variation in Pleistocene Australians. Mankind 8:85–91.
1972   Recent Discoveries of Fossil Man in Australia. Australia Natural History (June issue): 191–195.
1975   Kow Swamp and Lake Mungo. Ph.D. dissertation, University of Sydney.

Thorne, Alan G., and P. G. Macumber
1972   Discoveries of Late Pleistocene Man at Kow Swamp, Australia. Nature 238:316–319.

Thorne, Alan G., and Milford H. Wolpoff
1981   Regional Continuity in Australasian Pleistocene Hominid Evolution. American Journal of Physical Anthropology 55:337–349.

Tobias, Phillip V.
1965a   New Discoveries in Tanganyika: Their Bearing on Hominid Evolution. Current Anthropology 6:391–411.
1965b   The Early *Australopithecus* and *Homo* from Tanzania. Anthropologie (Prague) 3:43–48.
1967   Olduvai Gorge, vol. 2: The Cranium and Maxillary Dentition of *Australopithecus* (*Zinjanthropus*) *boisei*. Cambridge: Cambridge University Press.
1971   The Brain in Hominid Evolution. New York: Columbia University Press.
1972   "Dished Faces," Brain Size and Early Hominids. Nature 239:468–469.
1974   The Taung Skull Revisited. Natural History 83:38–43.
1984   The Child from Taung. Science 84 5:99–100.
1985   The Former Taung Cave System in the Light of Contemporary Reports and Its Bearing on the Skull's Provenance: Early Deterrents to the Acceptance of *Australopithecus*. In (Phillip V. Tobias, editor) Hominid Evolution: Past, Present and Future, pp. 25–40. New York: Alan R. Liss.
1991   Olduvai Gorge, vol. 4: The Skulls, Endocasts and Teeth of *Homo habilis*. Cambridge: Cambridge University Press.

Tobias, Phillip V., and Dean Falk
   1988   Evidence for a Dual Pattern of Cranial Venous Sinuses on the Endocranial Cast of Taung (*Australopithecus africanus*). American Journal of Physical Anthropology 76:309–312.

Tobias, P. V., and G. H. R. von Koenigswald
   1964   A Comparison between the Olduvai Hominines and Those of Java and Some Implications for Hominid Phylogeny. Nature 204:515–518.

Trigger, Bruce G.
   1976   Nubia under the Pharaohs. Boulder, CO: Westview Press.

Trinkaus, Erik
   1975   The Neandertals from Krapina, Northern Yugoslavia: An Inventory of the Lower Limb Remains. Zeitschrift für Morphologie und Anthropologie 67:44–59.
   1978   Dental Remains from the Shanidar Adult Neanderthals. Journal of Human Evolution 7:369–382.
   1982a   Artificial Cranial Deformation of the Shanidar 1 and 5 Neandertals. Current Anthropology 23:198–199.
   1982b   A History of *Homo erectus* and *Homo sapiens* Paleontology in America. In (Frank Spencer, editor) A History of American Physical Anthropology, 1930–1980, pp. 261–280. New York: Academic Press.
   1983   The Shanidar Neandertals. New York: Academic Press.
   1984   Western Asia. In (Fred H. Smith and Frank Spencer, editors) The Origins of Modern Humans: A World Survey of the Fossil Evidence, pp. 251–293. New York: Alan R. Liss.
   1985   Pathology and Posture of the La Chapelle-aux-Saints Neandertal. American Journal of Physical Anthropology 67:19–41.
   1986   The Neandertals and Modern Human Origins. Annual Review of Anthropology 15:193–218.

Trinkaus, Erik, and Marjorie LeMay
   1982   Occipital Bunning among Later Pleistocene Hominids. American Journal of Physical Anthropology 57:27–35.

Trinkaus, Erik, and M. R. Zimmerman
   1982   Trauma among the Shanidar Neandertals. American Journal of Physical Anthropology 57:61–76.

Turner, C. G., II, and J. Bird
   1981   Dentition of Chilean Populations and Peopling of the Americas. Science 212:1053–1055.

Tuttle, Russell H.
   1988   What's New in African Paleoanthropology? Annual Review of Anthropology 17:391–426.

Tyler, D.E., S. Sartono, and G. S. Krantz
   1994   A New *Homo erectus* Skull from Sangiran, Java. American Journal of Physical Anthropology, Supplement 18:199.

Ullrich, H.
   1986   Manipulations on Human Corpses, Mortuary Practice and Burial Rites in Palaeolithic Times. In (Vladimir V. Novotny and Alena Mizerova, editors) Fossil Man: New Facts—New Ideas, pp. 227–236. Brno: Anthropos Institut-Moravian Museum. (Special issue of Anthropos 23).

Valladas, H., J. L. Reyss, J. L. Joron, G. Valladas, O. Bar-Yosef, and B. Vandermeersch
   1988   Thermoluminescence Dating of Mousterian "Proto-Cro-Magnon" Remains from Israel and the Origin of Modern Man. Nature 331:614–616.

Vallois, H., and G. Billy
   1965   Nouvelles Recherches sur les Hommes Fossiles de l'Abri de Cro-Magnon. L'Anthropologie 69:47–74, 249–272.

Vallois, H., and B. Vandermeersch
   1972   Le Crane Moustérien de Qafzeh (*Homo* VI). L'Anthropologie 76:71–96.

Vandermeersch, B.
   1972   Récentes Découvertes de Squelettes Humains à Qafzeh (Israel): Essai d'Interprétation. In (F. Bordes, editor) The Origin of *Homo sapiens*, pp. 49–53. Paris: Unesco.
   1981   Les Hommes Fossiles de Qafzeh (Israel). Paris: Centre National de la Recherche Scientifique.

Vogel, John C.
   1985   Further Attempts at Dating the Taung Tufas. In (Phillip V. Tobias, editor) Hominid Evolution: Past, Present and Future, pp. 189–194. New York: Alan R. Liss.

Von Bonin, G.
   1935   European Races of the Upper Paleolithic. Human Biology 7:196–221.

Walker, Alan
   1981   The Koobi Fora Hominids and their Bearing on the Origins of the Genus *Homo*. In (Becky A. Sigmon and Jerome S. Cybulski, editors) *Homo erectus*: Papers in Honor of Davidson Black, pp. 193–216. Toronto: University of Toronto Press.

1993 The Origin of the Genus *Homo*. In (D. Tab Rasmussen, editor) The Origin and Evolution of Humans and Humanness, pp. 29–48. Boston: Jones and Bartlett Publishers.

Walker, Alan, Dean Falk, Richard Smith, and Martin Pickford
1983 The Skull of *Proconsul africanus*: Reconstruction and Cranial Capacity. Nature 305:525–527.

Walker, Alan, and Richard E. F. Leakey
1978 The Hominids of East Turkana. Scientific American 239:54–66.
1984 New Fossil Primates from the Lower Miocene Site of Buluk, N. Kenya. American Journal of Physical Anthropology 63:232.
1988 The Evolution of *Australopithecus boisei*. In (Frederick E. Grine, editor) Evolutionary History of the "Robust" Australopithecines, pp. 247–248. New York: Aldine de Gruyter.

Walker, A., and R. Leakey (editors)
1993 The Nariokotome *Homo erectus* Skeleton. Cambridge: Harvard University Press.

Walker, A., R. E. Leakey, J. M. Harris, and F. H. Brown
1986 2.5-Myr *Australopithecus boisei* from West of Lake Turkana, Kenya. Nature 322:517–522.

Walker, A. C., and M. Pickford
1983 New Postcranial Fossils of *Proconsul africanus* and *Proconsul nyanzae*. In (R. L. Ciochon and R. Corruccini, editors) New Interpretations of Ape and Human Ancestry, pp. 325–352. New York: Plenum Press.

Walker, Alan, and Mark Teaford
1989 The Hunt for *Proconsul*. Scientific American 261:76–82.

Walker, A., M. F. Teaford, and R. E. Leakey
1986 New Information Concerning the R114 *Proconsul* Site, Rusinga Island, Kenya. In (J. G. Else and P. C. Lee, editors) Primate Evolution, pp. 143–149. New York: Cambridge University Press.

Walker, A., M. F. Teaford, L. Martin, and P. Andrews
1993 A New Species of *Proconsul* from the Early Miocene of Rusinga/Mfangano Islands, Kenya. Journal of Human Evolution 25:43–56.

Ward, Carol V.
1993 Torso Morphology and Locomotion in *Proconsul nyanzae*. American Journal of Physical Anthropology 92:291–328.

Ward, Carol V., Alan Walker, and Mark F. Teaford
1991 *Proconsul* Did Not Have a Tail. Journal of Human Evolution 21:215–220.

Ward, C. V., A. Walker, M. F. Teaford, and I. Odhiambo
1993 Partial Skeleton of *Proconsul nyanzae* from Mfangano Island, Kenya. American Journal of Physical Anthropology 90:77–111.

Ward, Ryk, and Chris Stringer
1997 A Molecular Handle on the Neanderthals. Nature 388:225–226.

Ward, S. C., and B. Brown
1986 The Facial Skeleton of *Sivapithecus indicus*. In (D. R. Swindler and J. Erwin, editors) Comparative Primate Biology, vol. 1: Systematics, Evolution, and Anatomy, pp. 413–452. New York: Alan R. Liss.

Ward, S. C., and W. H. Kimbel
1983 Subnasal Alveolar Morphology and the Systematic Position of *Sivapithecus*. American Journal of Physical Anthropology 61:157–171.

Ward, S. C., and D. Pilbeam
1983 Maxillofacial Morphology of Miocene Hominoids from Africa and Indo-Pakistan. In (R. L. Ciochon and R. Corruccini, editors) New Interpretations of Ape and Human Ancestry, pp. 211–238. New York: Plenum Press.

Washburn, S. L., and Ruth Moore
1980 Ape into Human, a Study of Human Evolution, second edition. Boston: Little, Brown and Co.

Weidenreich, Franz
1935 The *Sinanthropus* Population of Choukoutien (Locality 1) with a Preliminary Note on New Discoveries. Bulletin of the Geological Society of China 14:427–468.
1936a Observations on the Form and the Proportions of the Endocranial Casts of *Sinanthropus pekinensis*, Other Hominids, and the Great Apes: A Comparative Study of Brain Size. Palaeontologica Sinica (series D) 7.
1936b The Mandible of *Sinanthropus pekinensis*: A Comparative Study. Palaeontologica Sinica (series D) 7(3): 1–162.
1937 The Dentition of *Sinanthropus pekinensis*: A Comparative Odontography of the Hominids. Palaeontologica Sinica (Series D) 1 (whole series 101): 1–180.
1939 The Duration of Life of Fossil Man in China and the Pathological Lesions Found in His Skeleton. Chinese Medical Journal 55:34–44.
1940 Man or Ape? Natural History 45:32–37.

1943   The Skull of *Sinanthropus pekinensis*. Palaeontologica Sinica (new series D) 10.

1945   Giant Early Man from Java and South China. Anthropological Papers of the American Museum of Natural History 40, part 1.

1946a   Apes, Giants, and Man. Chicago: University of Chicago Press.

1946b   Report on the Latest Discoveries of Early Man in the Far East. Experienta 2:265–272.

1951   Morphology of Solo Man. Anthropological Papers of the American Museum of Natural History 43, part 3.

Weigelt, Johannes

1989   Recent Vertebrate Carcasses and Their Paleobiological Implications. Chicago: University of Chicago Press.

Weiner, J. S.

1971   The Natural History of Man. New York: Universe Books.

Weiner, J. S., and B. G. Campbell

1964   The Taxonomic Status of the Swanscombe Skull. In (C. D. Ovey, editor) The Swanscombe Skull. Occasional Papers of the Royal Anthropological Institute 20:175–209.

Weinert, H.

1936   Der Urmenschenschädel von Steinheim. Zeitschrift für Morphologie und Anthropologie 35:463–518.

Wendorf, Fred, Romuald Schild, Angela E. Close, Gordon C. Hillman, Achilles Gautier, Wim Van Neer, D. J. Donahue, A. J. T. Hull, and T. W. Linick

1988   New Radiocarbon Dates and Late Palaeolithic Diet at Wadi Kubbaniya, Egypt. Antiquity 62:279–283.

White, T. D.

1975   Geomorphology to Paleoecology: *Gigantopithecus* Reappraised. Journal of Human Evolution 4:219–233.

1977   New Fossil Hominids from Laetoli, Tanzania. American Journal of Physical Anthropology 46:197–230.

1980   Additional Fossil Hominids from Laetoli, Tanzania: 1976–1979 Specimens. American Journal of Physical Anthropology 53:487–504.

1984   Pliocene Hominids from the Middle Awash, Ethiopia. In (Peter Andrews and Jens Lorenz Franzen, editors) The Early Evolution of Man, with Special Emphasis on Southeast Asia and Africa, pp. 57–68. Courier Forschungsinstitut Senckenberg 69.

1985   Acheulian Man in Ethiopia's Middle Awash Valley: The Implications of Cutmarks on the Bodo Cranium. Kroon Memorial Lecture. Haarlem, Netherlands: Enschede en Zonen.

1986   Cut Marks on the Bodo Cranium: A Case of Prehistoric Defleshing. American Journal of Physical Anthropology 69: 503–509.

1991   Human Osteology. San Diego: Academic Press.

White, Tim D., and Donald C. Johanson

1982   Pliocene Hominid Mandibles from the Hadar Formation, Ethiopia: 1974–1977 Collections. American Journal of Physical Anthropology 57:501–544.

White, T. D., D. C. Johanson, and W. H. Kimbel

1981   *Australopithecus africanus*: Its Phyletic Position Reconsidered. South African Journal of Science 77:445–470.

White, Tim D., and Gen Suwa

1987   Hominid Footprints at Laetoli: Facts and Interpretations. American Journal of Physical Anthropology 72:485–514.

White, Tim D., Gen Suwa, and Berhane Asfaw

1994   *Australopithecus ramidus*, a New Species of Early Hominid from Aramis, Ethiopia. Nature 371:306–312.

1995   Corrigendum: *Australopithecus ramidus*, a New Species of Early Hominid from Aramis, Ethiopia. Nature 375:88.

White, T. D., Gen Suwa, William K. Hart, Robert C. Walter, Giday WoldeGabriel, Jean de Heinzelin, J. Desmond Clark, Berhane Asfaw, and Elisabeth Vrba

1993   New Discoveries of *Australopithecus* at Maka in Ethiopia. Nature 366:261–265.

White, Tim D., and Nicholas Toth

1991   The Cultural Significance of Grotta Guattari Reconsidered: 2. The Question of Ritual Cannibalism at Grotta Guattari. Current Anthropology 32:118–124.

WoldeGabriel, Giday, Tim D. White, Gen Suwa, Paul Renne, Jean de Heinzelin, William K. Hart, and Grant Heiken

1994   Ecological and Temporal Placement of Early Pliocene Hominids at Aramis, Ethiopia. Nature 371:330–333.

Wolpoff, Milford H.

1970   Evidence for Multiple Hominid Taxa at Swartkrans. American Anthropologist 72:576–607.

1971   Is the New Composite Cranium from Swartkrans a Small Robust Australopithecine? Nature 230:398–401.

1979   The Krapina Dental Remains. American Journal of Physical Anthropology 50:67–113.

1982   *Ramapithecus* and Hominid Origins. Current Anthropology 23:501–522.

1983a   Lucy's Lower Limbs: Long Enough for Lucy to be Fully Bipedal? Nature 304:59–61.

1983b   Lucy's Little Legs. Journal of Human Evolution 12:443–453.

1983c   *Ramapithecus* and Human Origins: An Anthropologist's Perspective of Changing Interpretations. In (R. L. Ciochon and R. Corruccini, editors) New Interpretations of Ape and Human Ancestry, pp. 651–676. New York: Plenum Press.

1996   Human Evolution. New York: McGraw-Hill, College Custom Series.

1998   Paleoanthropology, second edition. New York: McGraw-Hill.

Wolpoff, Milford H., Janet M. Monge, and Michelle Lampl

1988   Was Taung Human or an Ape? Nature 335:501.

Wolpoff, Milford H., Wu Xin Zhi, and Alan G. Thorne

1984   Modern *Homo sapiens* Origins: A General Theory of Hominid Evolution Involving the Fossil Evidence from East Asia. In (Fred H. Smith and Frank Spencer, editors) The Origins of Modern Humans: A World Survey of the Fossil Evidence, pp. 411–483. New York: Alan R. Liss.

Wood, B. A.

1976   Remains Attributable to *Homo* in the East Rudolf Succession. In (Yves Coppens, F. Clark Howell, Glynn Ll. Isaac, and Richard E. F. Leakey, editors) Earliest Man and Environments in the Lake Rudolf Basin, pp. 409–506. Chicago: University of Chicago Press.

1991   Koobi Fora Research Project, vol. 4: Hominid Cranial Remains. Oxford: Clarendon Press.

Woodward, Arthur Smith

1921   A New Cave Man from Rhodesia, South Africa. Nature 108:371–372.

Wu Rukang

1983   Hominid Fossils from China and Their Bearing on Human Evolution. Canadian Journal of Anthropology 3:207–214.

1985   New Chinese *Homo erectus* and Recent Work at Zhoukoudian. In (Eric Delson, editor) Ancestors: The Hard Evidence, pp. 245–248. New York: Alan R. Liss.

1987   A Revision of the Classification of the Lufeng Great Apes. Acta Anthropologica Sinica 6:265–271.

Wu Rukang, and Dong Xingren

1982   Preliminary Study of *Homo erectus* Remains from Hexian, Anhui. Acta Anthropologica Sinica 1:2–13.

1983   Des Fossiles d'*Homo erectus* Découverts en Chine. L'Anthropologie 87:177–183.

1985   *Homo erectus* in China. In (Wu Rukang and John W. Olsen, editors) Palaeoanthropology and Palaeolithic Archaeology in the People's Republic of China, pp. 79–89. Orlando: Academic Press.

Wu Rukang, Han Defen, Xu Quinghua, Lu Qingwu, Pan Yuerong, Zhang Xingyong, Zheng Liang, and Xiao Minghua

1981   *Ramapithecus* Skulls Found First Time in the World. Kexue Tongbao 26:1018–1021.

Wu Rukang, and Lin Shenglong

1983   Peking Man. Scientific American 248:86–94.

Wu Rukang, and Charles Oxnard

1983   *Ramapithecus* and *Sivapithecus* from China: Some Implications for Higher Primate Evolution. American Journal of Primatology 5:313–344.

Wu Rukang, and Xu Quinghua

1985   *Ramapithecus* and *Sivapithecus* from Lufeng, China. In (Wu Rukang and John W. Olsen, editors) Palaeoanthropology and Palaeolithic Archaeology in the People's Republic of China, pp. 53–68. Orlando: Academic Press.

Wu Rukang, Xu Qinghu, and Lu Qingwu

1983   Morphological Features of *Ramapithecus* and *Sivapithecus* and Their Phylogenetic Relationship—Morphology and Comparison of the Crania. Acta Anthropologica Sinica 2:1–10. (Translation: Yearbook of Physical Anthropology 27:31–40.)

Wu Xinzhi

1981   A Well-Preserved Cranium of an Archaic Type of Early *Homo sapiens* from Dali, China. Scientia Sinica 24:530–541.

Wu Xinzhi, and Frank Poirier

1995   Human Evolution in China: A Metric Description of the Fossils and a Review of the Sites. New York: Oxford University Press.

Wu Xinzhi, and Wu Maolin

1985   Early *Homo sapiens* in China. In (Wu Rukang and John W. Olsen, editors) Palaeoanthropology and Palaeolithic Archaeology in the People's Republic of China, pp. 91–106. Orlando: Academic Press.

Wymer, J.

1955   A Further Fragment of the Swanscombe Skull. Nature 176:426–427.

Xu Qinghua, and Lu Qingwu

1979   The Mandibles of *Ramapithecus* and *Sivapithecus* from Lufeng, Yunnan. Vertebrata Palasiatica 17: 1–13. (Translation in Yearbook of Physical Anthropology 27:15–23.)

1980   The Lufeng Ape Skull and Its Significance. China Reconstructs 29:56–57.

Xu Qinghua, Lu Qingwu, Pan Yuerong, Qi Guoqin, Zhang Xingyong, and Zheng Liang
    1978   The Mandible of *Ramapithecus lufengensis*. Kexue Tongbao 9:546, 554–556. (Translation: Yearbook of Physical Anthropology 27:2–6.)

Zhang Yinyun
    1982   Variability and Evolutionary Trends in Tooth Size of *Gigantopithecus blacki*. American Journal of Physical Anthropology 59:21–32.
    1985   *Gigantopithecus* and *"Australopithecus"* in China. In (Wu Rukang and John W. Olsen, editors) Palaeoanthropology and Palaeolithic Archaeology in the People's Republic of China, pp. 69–78. Orlando: Academic Press.